THE OP
MATHE
AN INTI
MA290:
TICS

BLOCK 3   THE SEVENTEENTH AND EIGHTEENTH
CENTURIES

UNIT 11

MATHEMATICAL PHYSICS AND THE SYSTEM OF THE WORLD

PREPARED BY JEREMY GRAY FOR THE COURSE TEAM

THE OPEN UNIVERSITY

CONTENTS

11.0	**Introduction**	**3**
11.1	**The Creation of Newton's *Principia***	**3**
11.2	**The Content of the *Principia***	**9**
11.3	**The Impact of the *Principia***	**16**
11.4	**Rational Mechanics**	**23**
11.5	**The Analysis of Nature**	**28**
	Further Reading	**34**

This unit forms part of an Open University course. The set book for the course, to which reference is made as **SB**, is:

John Fauvel and Jeremy Gray (editors), *History of Mathematics: A Reader*, Macmillan 1987.

Acknowledgements

Grateful acknowledgement is made to the following sources for material used in this unit: *Figure 1*, E. J. Aiton, *The Vortex Theory of Planetary Motions*, (Macdonald, 1972); *Figure 2*, R. S. Westfall, *Never at Rest* (Cambridge University Press, 1980); *Figure 3*, the Master and Fellows of Magdalene College; *Figure 4*, Ronan Picture Library; *Figures 5, 9, 10* and *12*, Keele University, Turner Collection; *Figure 7*, Photographie Giraudon; *Figure 8*, Snark International, Paris; *Figure 11*, T. Perl, *Math Equals* (Addison-Wesley, 1978); *Figure 14*, National Maritime Museum; *Figures 16, 17* and *21*, University of Basle Library; *Figure 20*, the Petworth Estate; *Figure 27*, T. L. Hankins, *Science and the Enlightenment* (Cambridge University Press, 1985).

The Open University, Walton Hall, Milton Keynes.

First published 1987. Reprinted 1990, 1995.

Designed by the Graphic Design Group of the Open University.

Typeset in Great Britain by Santype International Ltd, Salisbury.

Printed in Great Britain by BPC Wheatons Ltd, Exeter.

ISBN 0 335 14255 9

This text forms part of the correspondence element of an Open University Second Level Course.

For general availability of supporting material referred to in this text, please write to Open University Educational Enterprises Limited, 12 Cofferidge Close, Stony Stratford, Milton Keynes, MK11 1BY, Great Britain.

Further information on Open University courses may be obtained from The Admissions Office, The Open University, P.O. Box 48, Milton Keynes, MK7 6AB.

11.0 INTRODUCTION

In this unit we consider the historical development of the mathematical study of the natural world, concentrating in particular on *celestial mechanics*, the mathematics of planetary motion. This formed the most significant application of mathematics throughout the seventeenth and eighteenth centuries. In the first section we look at what was believed by the 1670s (not all of it correctly) on the basis of the work of Galileo, Kepler, Descartes, and Huygens. This body of theory formed the basis upon which Newton was to create his *Principia*, and we look at how he came to do this and the circumstances in which this work was brought to publication.

The *Principia* itself is the focus of the second section, where we highlight its structure—part mathematics, part analysis of physical hypotheses of the causes of motion of the planets. Perceiving this structural aspect is crucial to our attempt to understand the reception of the book, which we consider in Section 3. Two problems dominated the mid-eighteenth-century debate about Newton's ideas: the shape of the Earth and the motion of the moon. We examine this debate, which left Newtonianism securely established as the paradigm of a mathematical science.

In Section 4 we try to pin down more precisely what mathematical science included and what it did not, before looking at some of the reasons advanced as to how it was that mathematics itself could play such a central role in science. In the final section we reverse this emphasis and look at the kind of mathematics which this involvement in science brought forth, namely rational mechanics or the mathematical analysis of nature.

11.1 THE CREATION OF NEWTON'S *PRINCIPIA*

To understand the *Principia* and its reception, it is helpful to begin by looking at the context in which Newton proceeded, for the book was the culmination of twenty years during which he struggled to master both mechanics and astronomy. Newton had four major seventeenth-century predecessors in his studies: Galileo, Kepler, Descartes, and Huygens. From Galileo's *Discourses concerning the Two Chief World Systems*, translated into English by Salusbury in 1661, Newton seems to have learned the laws governing the descent of bodies under earthly gravity and also about inertia, a topic we discuss below. But Newton never read the *Discourse concerning the Two New Sciences*, Galileo's last work in which (as we saw in *Unit 7*, Section 2) his ideas about motion were given their most thorough mathematical exposition. From Kepler's work, Newton could have learned the most careful and profound description of the motion of the planets. But he never read any of Kepler's own accounts, and professional astronomers were divided in their assessment of the planetary laws. They agreed in the main with Kepler's first law (that planets travel in ellipses with the sun at one focus) and with the third (which relates the period of the orbit to its mean radius). But astronomers had little use for the second law (that the line joining any planet to the sun sweeps out equal areas in equal times), because they found it impossible to calculate with. As a result, the books Newton read tended to state this law—if they mentioned it at all—only to reject it in favour of one better adapted to calculation. Moreover, Kepler's physical explanation of the motion of the planets, a kind of magnetic force which emanated from the sun and somehow pushed the planets round, never won any support. Indeed, as we also saw in *Unit 7*, the works of Kepler and Galileo were curiously unconnected, and it was to be others who tried to produce a theory of planetary motion which could explain on more convincing mechanical grounds why the planets moved as Kepler said they did.

The most successful proponent of such a theory was Descartes, who argued in his *Principles of Philosophy* (1644), that the Universe was full of invisible little particles which swept round the sun like a cosmic whirlpool or *vortex*. The action of this vortex was to drive the planets round, and the result was that all the planets should lie in the same plane and orbit the sun in the same direction. This theory had three things to recommend it: it explained why the orbits of the planets do all lie more or less in the same plane; it explained why they all go round the sun in the same direction; and it explained this by invoking an intuitively simple mechanism for celestial motion—collision. For these reasons, his 'vortex theory' was widely accepted.

Or *Principia Philosophiae*.

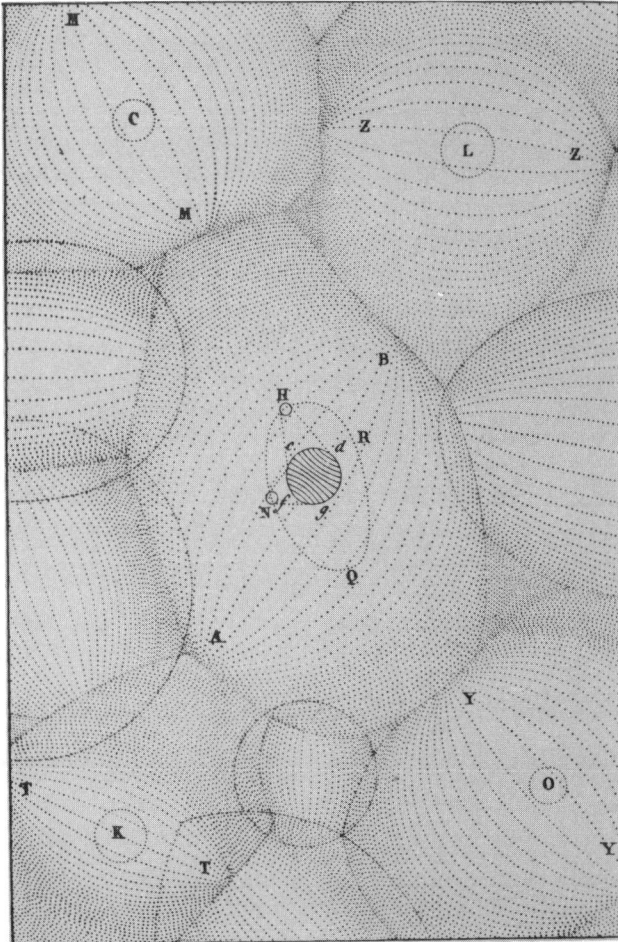

Figure 1 Cartesian vortices

Christiaan Huygens was later to say that when he first read Descartes' book—he was then fifteen or sixteen—he thought everything was 'splendid' but that later he 'recovered a good deal from the infatuation I had for it.' Indeed, as a theory it left much to be desired. In particular, it did not account for the motion of the planets with the precision required by Kepler's laws. Nor did Descartes' detailed account of collisions stand up to much scrutiny, and the book is remarkable in its disdain for experiment and observation. It is, rather, an *a priori* physics based on some ideas which were clear and immediate to the mind of its author. For this reason Huygens, like Pascal, called Descartes' book a 'romance', full of conjectures and fictions which came to be accepted as truths because of their intrinsic charm. As we shall see, this book both provided Newton with his first instruction in the theory of motion, and was to be decisively attacked by him in *Principia*.

Quoted in A. R. Hall, *The Revolution in Science, 1500–1750* (Longman, 1983) p. 295.

Huygens' own ideas were more precise. For example, he gave a careful account of circular motion, based on the idea that there is what he called a centrifugal force. He imagined a body being whirled rapidly around a fixed, central point like a stone in a sling, and found a quantitative relationship between the radius of the circle and the size of the centrifugal force. But Huygens was unhappy with the idea of a force as a fundamental principle; he remained all his life an enthusiast for the Cartesian idea of collisions, as the kind of mechanistic concept that made it possible ultimately to *explain* phenomena.

The centrifugal force is what he supposed would drive the stone away if the sling were to break—see Box 1.

Box 1 Centripetal and centrifugal forces

According to Newton, all bodies attract one another by a force, which he called *gravity*. The effect of a force acting on a body is to cause it to accelerate in the direction of the force—so Newton's force is an example of a *vectorial quantity* (or *vector*), one having both a size and a direction.

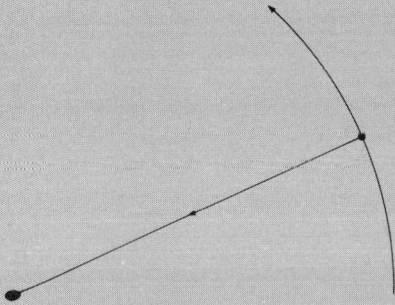

As a simplification of the situation in which two bodies attract one another, consider the case in which one body is so large that it remains at rest. (This would be a reasonable first approximation to the situation where the bodies are the sun and a planet.) Then the smaller body is attracted to the larger one, and this force therefore always pulls it towards the same point. This is an example of what Newton called a *centripetal* force, meaning a force directed towards a fixed point, which will, in some intuitive sense, be the *centre* of the orbit.

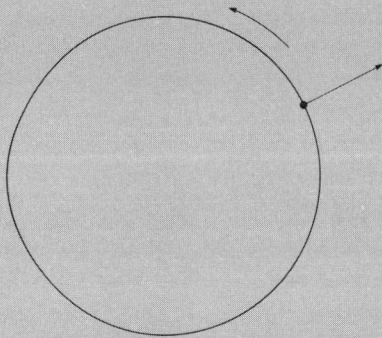

This differs from Huygens' idea of a force which pushes a body away from the centre, which Huygens had called a *centrifugal* force, meaning a force causing the body to flee from the centre. This is the force you experience if you are swung round, say at the end of a rope, or in a rotating drum. The word is used today to describe such things as a spin dryer, where we say the water is driven out by the centrifugal force generated by the rotating drum.

From this brief survey, it is clear that the task facing Newton was still immense. A *causal* theory of motion capable of yielding accurate quantitative descriptions was still lacking, even for terrestrial physics, and such a theory covering both small objects near the Earth and the motion of the planets must have seemed still further out of reach. Kepler's idea of a solar force was rejected, and Descartes' more plausible ideas had not been made sufficiently precise. Let us see how Newton came to create just such a unified theory.

In England, the theory of celestial mechanics was much debated in the late 1670s by fellows of the Royal Society such as Hooke, Halley, and Wren; and the debate was quickened by the magnificent comet of 1680, which could be seen in daytime for weeks, and at its largest stretched over nearly a third of the visible sky, making it at least 60 million miles long. But the final impulse which set the *Principia* in motion was a visit Edmond Halley paid to Newton in August 1684. By January of that year Hooke, Wren, and Halley had, by combining Kepler's third law with Huygens' formula for centrifugal force, come to the view that the sun attracts all the planets

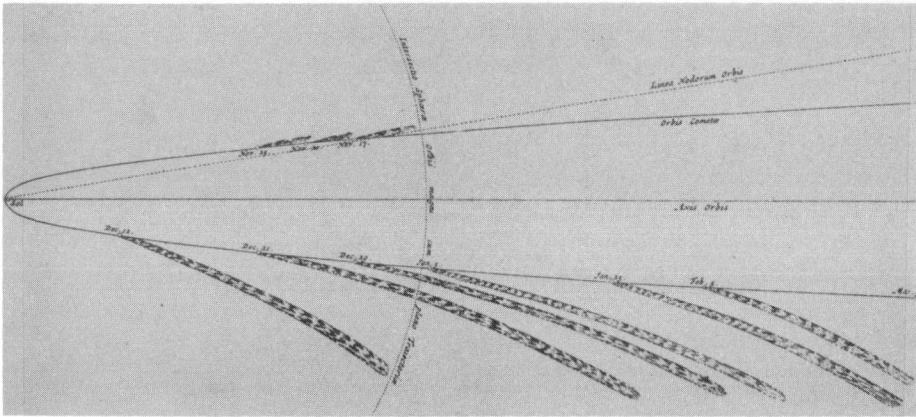

Figure 2 The 'true representation of the orbit' of the great comet of 1680, from Newton's *Principia*

according to an inverse square law. However, none of them could derive all of Kepler's laws from dynamical principles, and Halley decided to travel to Cambridge and ask for Newton's help. Much later, in 1722, Abraham de Moivre wrote down this account of how Newton remembered the visit:

> D^r [Halley] asked him what he thought the Curve would be that would be described by the Planets supposing the force of attraction towards the Sun to be reciprocal to the square of their distance from it. Sr Isaac replied immediately that it would be an Ellipsis, the Doctor struck with joy & amazement asked him how he knew it, why saith he I have calculated it, whereupon Dr Halley asked him for his calculation without any farther delay, Sr Isaac looked among his papers but could not find it, but he promised him to renew it, & then to send it him.

Newton set himself to re-derive the solution, and sent to the Royal Society in November a nine-page tract called *De Motu Corporum* ('On the motion of bodies'). In this he showed how an elliptical orbit and a centripetal force together imply an inverse square law for the force; he also sketched the converse, thus answering Halley's question, to wit, that an inverse square law for a centripetal force implies that the orbit is a conic section. The tract caused a great stir, and was eagerly read by members of the Royal Society. Halley went back to Cambridge in November, to urge Newton perhaps to publish *De Motu Corporum*, perhaps to amplify it. But Newton had already allowed himself to become totally immersed in the problems of physics and celestial mechanics. Until the spring of 1686 he seems to have thought of nothing else. He wrote repeatedly to Flamsteed for the most accurate data, and continued to lecture at Cambridge, but otherwise he disappeared entirely from

A theory of motion is called *dynamical* if it invokes a concept of force—any description of motion which does not, and merely considers speeds, is called *kinematical*.

R. S. Westfall, *Never at Rest* (Cambridge University Press, 1980) p. 403.

John Flamsteed (1646–1719) was the first Astronomer Royal. An observational astronomer of unrivalled precision, his later life was marred by disputes with Newton and Halley over publication of his results.

PROSPECTUS INTRA CAMERAM STELLATAM

Figure 3 Interior of the Royal Observatory, Greenwich, c. 1676

intellectual life for two years. In so doing he returned to the habits of his first adult years, and also to his original questions; but this time he was to publish his results and give the world evidence of his brilliance.

Box 2   Halley's Comet—and others

The significance of comets in the seventeenth century arises from the fact that it was beginning to be realised that they were true celestial objects. Galileo, for example, was somewhat behind the times in holding to the old, Aristotelian belief that comets were phenomena of the upper atmosphere. Once they were recognised as being really up in the heavens—beyond the moon—they raised some interesting questions. Along what paths, for example, did they travel? Most people thought, like Kepler, that they travelled along straight paths, either into or out of the sun, although not at a uniform speed. In 1680, Flamsteed discussed the so-called 'great comet' in correspondence with Newton, and raised the idea that the comet orbited the sun and changed its direction as it made its closest approach. But Newton did not accept Flamsteed's ideas until 1682, when he made some observations of another comet (the one we now call Halley's)—which tells us that in 1680 Newton was some way from regarding gravitation as universal. But when writing *De Motu Corporum*, he observed quite excitedly that the inverse square law 'allowed [one] to define the orbit of comets and thereby their periods of revolution'.

D. T. Whiteside (ed.) *The Mathematical Papers of Isaac Newton*, vol. VI (Cambridge University Press, 1971) p. 57.

Newton was, in fact, the first to compute the orbit of a comet (he took the great comet of 1680–1) and his treatment of this problem in the *Principia* is one of the great successes of the work. Moreover, once the orbit of comets became better understood, it emerged that some (like the great comet) move around the sun in the opposite direction to the planets. This is called retrograde motion, and is a fact difficult to reconcile with Descartes' vortex theory.

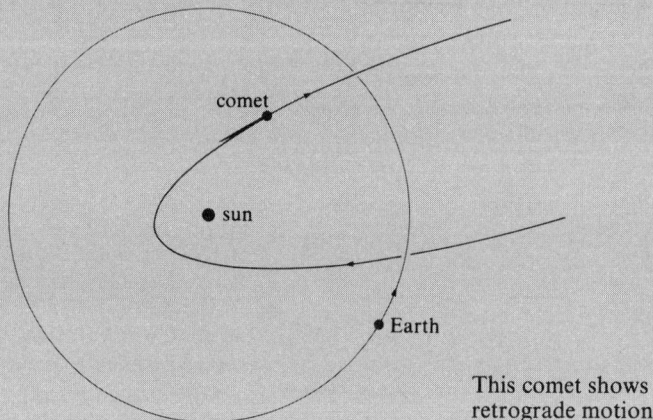

This comet shows retrograde motion

And Halley? A keen astronomer and a skilful mathematician, it was his achievement to read the existing literature in the light of the new, orbital theory of comets and, in 1705, to pronounce some of them regular visitors to our skies. In particular, he estimated that the comet of 1682 had a period of about 75 years. Accordingly he predicted its return in 1758, hoping that, if he were proved correct, 'candid posterity will not refuse to acknowledge that this was first discovered by an Englishman'.

Newton's first task was to produce a good theory of dynamics: what *is* a force acting on a body? This was not easy—indeed, in *De Motu Corporum* he had derived elliptical orbits on dynamically incorrect grounds. It took him six months to elucidate the correct notion of force to put into *Principia*, and to isolate the concept of *inertia* with which to express a body's tendency to keep on going with a uniform velocity if not acted on by an external force. Another problem Newton had to solve was this: if a theory is to be quantitatively accurate, the basic terms must be quantifiable—they must be able to have numbers attached to them—but how was *the quantity of motion* to be measured? This question had stumped Newton's

Figure 4   Edmond Halley (1656–1743)

contemporaries. Newton proposed *momentum*, the product of mass and velocity. In his second law he expressed the way in which force acts so as to cause a change in momentum, and in this way established the dynamical principles he needed. Please *look now* at the celebrated passage towards the beginning of *Principia* where he stated his axioms or laws of motion (**SB** 12.B2).

Question 1   What status in the *Principia* do you judge these laws to have—that is, does Newton appear to prove them? Secondly, how strongly mathematical do they appear to you to be?

Comment ──

These laws are not proved, as indeed the word 'axioms' implies. The text following the statement of each law is further explanation and elucidation of it, not justification. So these laws are foundations on which subsequent deductions are to be built, about the behaviour of moving bodies and bodies with forces acting on them.

Newton's laws are not visibly mathematical, or at any rate algebraic, in the sense that they do not have an appearance of equations, symbols and their manipulation. If you have studied modern applied mathematics you may be surprised at how far we are from modern conceptual formulations such as (for Law II) 'force = rate of change of momentum'. ■

Newton also had to decide what were the crucial astronomical ideas he had to explain on the basis of his theory of dynamics. He chose to go for all three of Kepler's laws. When you look at our summary of the *Principia* in Section 2 you will see how Newton's vast generalisation of Kepler's second (equi-area) law, prominently placed near the front of the book, was to play a vital role in the theory he presented. To state and prove the law in such generality may, indeed, be Newton's greatest contribution to dynamics. Certainly everyone before him had thought that Kepler's laws applied only to the planets and were hard to justify further—Newton showed that their explanation rested on quite general dynamical grounds. Of course, as Kepler had known and as Newton more-or-less proved, any real use of the equi-area law to describe or calculate an orbit must itself be approximate. Newton called the curve which measures the area of a sector of an ellipse 'geometrically irrational', which was his phrase for what Descartes had called mechanical curves. In the context of practical astronomy, Newton's remarks confirmed that astronomers would have to continue to settle for only approximate answers. While this may partially vindicate those who replaced Kepler's law with their own, it does nothing to diminish Newton's achievement. I. B. Cohen has observed that

> it was an unusual and a very daring step to erect an astronomical system encompassing Kepler's three laws, as Newton did. Following the imaginative leap forward that Newton made, in showing the physical meaning and conditions of mathematical generality or applicability of each of Kepler's laws, this whole set of three laws gained a real status in exact science.

I. B. Cohen, *The Newtonian Revolution* (Cambridge University Press, 1980) p. 229.

It is also worth noticing that the *Principia* takes it for granted that the solar system is heliocentric, because earlier drafts did not. Newton had written lengthy justifications for heliocentrism, but then suppressed them.

The circumstances in which the *Principia* came into the world are dramatic. In April 1686 the manuscript of Book I was presented to the Royal Society. Halley continued his role as the work's midwife, and successfully pursued his goal of ensuring that Newton's masterwork was printed and published in its entirety, against obstacles that would have exhausted a less forceful person. For the Royal Society, whose finances were somewhat delicately poised throughout these years, had just succeeded in nearly bankrupting itself through the publication of a very handsome, expensive and unsellable *History of Fishes*; there was certainly no money spare for printing Mr Newton's thoughts upon the Cosmos. Indeed the Royal Society was reduced to recompensing Halley—an employee of the Society—in copies of the *History of Fishes*, much as if the Open University were to survive the next round of financial cuts by paying its tutors in redundant course units. In the end Halley had to pay for the publication of *Principia* himself—fortunately the work was to make a profit, unlike the *History of Fishes*, so he did not lose by his generous gesture. But he had a greater problem to contend with: was the author going to allow his work to be published at all? The ever-touchy Newton had been roused to thunderous rage by learning that Robert Hooke was claiming not only priority in discovering the inverse square law, but also that Newton had taken it from him, and more besides. As the year went on, Newton announced to Halley that he would withdraw Book III altogether, if the reward of his labours was to be this irritation of false claims. At length Halley, a model of saintly tact and diplomacy, soothed Newton; the printing went on its way.

The moral of this tiresome squabble points up nicely the significance of *Principia*. Hooke's claim was actually true, in the very limited sense that he did consider an inverse square law earlier than Newton—as others had done, too—and anticipated Newton also in the realisation that a force must be acting on a body in orbit to deflect it constantly out of a straight line. Hooke was an ingenious man with bright, appropriate ideas. But unfortunately he was quite unable to distinguish between having an isolated, well-informed idea and the massive systematic endeavour of deducing a whole System of the World on dynamical principles.

At last, on 5 July 1687, Halley's task was done: the *Principia* was a printed book. With its publication, Newton's status as a mathematician, and the nature of mathematical physics, were changed utterly. Perhaps it was Newton, agreeably taking himself less seriously than usual, who spread the story that he passed a student in the Cambridge street who said 'there goes the man that writt a book that neither he nor any body else understands'.

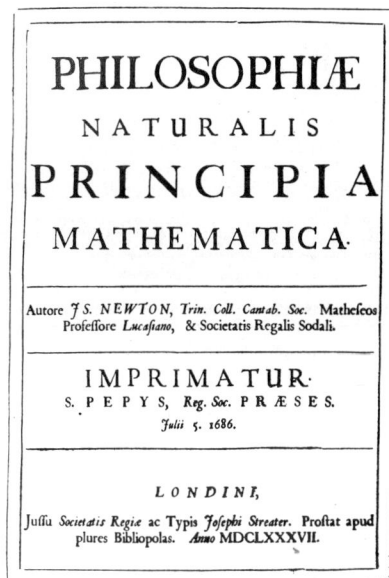

PHILOSOPHIÆ
NATURALIS
PRINCIPIA
MATHEMATICA.

Autore *JS. NEWTON*, *Trin. Coll. Cantab. Soc.* Mathefeos
Profeffore *Lucafiano*, & Societatis Regalis Sodali.

IMPRIMATUR·
S. P E P Y S, *Reg. Soc.* P R Æ S E S.
Julii 5. 1686.

LONDINI,
Juffu *Societatis Regiæ* ac Typis *Jofephi Streater.* Proftat apud
plures Bibliopolas. *Anno* MDCLXXXVII.

Figure 5 Title page to the first edition of Newton's *Principia*

Cited in Westfall, *Never at Rest*, p. 468.

11.2 THE CONTENT OF THE *PRINCIPIA*

Imagine that Newton's *Principia* has just come into your hands, one day in late 1687. You first of all turn over the 547 pages, written in scholarly Latin, to get an impression of what it contains. This book began to create a stir even before it was published. What is it actually about? The notes which follow provide an overview, a guide of sorts, with which to begin to answer that question. Read them over first as if you were skimming the book itself. Then we shall return to consider various passages of the *Principia* in more detail, and consult extracts from it, in order to get a more precise grasp of certain topics from this remarkable work.

Book I is preceded by definitions and axioms: definitions of quantity of matter; quantity of force; forces of various kinds; a distinction between relative and absolute motion; relative and absolute time. Then there are the three axioms or laws of motion you looked at above, and some elementary consequences of them.

Book I   The Motion of Bodies

This book opens with a long, careful, cumulative discussion of 'the method of first and last ratios of quantities', which is a geometrical study of curves and their tangents in the spirit in which Newton conducted his investigations of the calculus. There follows a lavish study of the motion of a point under a centripetal force (Newton's term, meaning a force directed towards a fixed point—see Box 1). Newton established that the line joining a fixed point to a moving one sweeps out equal areas in equal times if and only if the force on the moving point is directed towards the fixed point. (Strictly speaking the proof is valid only for infinitesimal intervals of time, but no-one noticed.) This is a very general result. The size of the force can depend in *any* way on the length of the radius line; the orbit can be *any* shape determined by the law, not just a circle or an ellipse. Numerous special cases are then worked out, including this one: if the moving body traverses a conic section under a centripetal force directed towards one focus, then the magnitude of the force is inversely proportional to the square of the distance. This, as Newton well knew, is the relevant case in astronomy, and he established the converse also: under an inverse square law, bodies move in curves which are conic sections having the centre of force as one focus.

Newton considered the special case of two bodies mutually attracting one another, and investigated various ways in which the size of this force can depend on the distance between them. If the size of the force is *inversely* proportional to the *square* of the distance, then the force is said to obey an *inverse square law*. This means that if the distance between the two bodies is doubled, then the size of the force is reduced to a quarter of what it was, and so on; so nearby bodies exert more of an influence than distant ones.

If you consider the special case raised in Box 1 of a large, stationary body and a small, moving one, then since the force on the moving body is always directed towards the fixed one, it is natural to wonder why they do not eventually collide, according to this theory of motion. In Hooke's view, this was somehow because the smaller body has a velocity directed away from the larger one (if it had not, they would certainly collide). One of Newton's achievements was to show on mathematical grounds that Hooke's insight was correct.

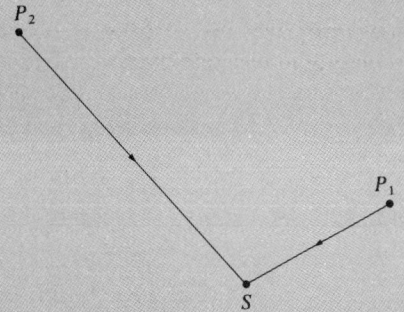

An inverse square law. Other things being equal, the distance from S to P_1 being half the distance SP_2, the force exerted on P_1 by S is four times the force on P_2

Now an astronomer needs to know of a planet not only what orbit it has as a whole, but where along that orbit it can be found at any particular time. Newton tackled this problem next, and showed in principle how to solve it using Kepler's equi-area law. He also showed that if planets traverse ellipses under the action of a force that obeys an inverse square law, then they necessarily obey Kepler's third ($\frac{3}{2}$ power) law. He even showed how to determine the orbit, in principle, given any centripetal law of force whatever.

We pass over a number of propositions including several concerning conic sections, and come to Newton's discussion of the attraction between solid bodies under an inverse square law. Here he established that a spherical shell exerts no force on a point inside it and attracts a point outside it in the same way as a point mass concentrated at the centre. This result, which surprised him as much as his contemporaries, enabled him to reduce the study of large spherical objects like planets to the study of points and centripetal forces, which he had already described.

The gravitational pull of a spherical shell

A remarkable result Newton obtained was that a thin spherical shell of matter of uniform density attracts bodies outside it exactly as would a point mass, situated at the centre of the shell and having the same mass as the shell. This means that a solid sphere, thought of as a nest of such shells, also attracts bodies outside it exactly as does a point mass. So in Newton's theory of gravity, large solid spheres may be replaced by *points* (of the same mass)—a considerable simplification in the theory. For much work in astronomy, the assumption that planets and the sun are spherical in shape is entirely reasonable.

There is no gravitational pull inside a spherical shell

To understand this result, consider a point, P, inside a spherical shell (and not necessarily at the centre, or there is nothing to prove!). Consider the attraction on it of a small piece of the surface, S, and the piece of the surface, S', opposite to S. Although one piece, say S, is nearer to P, it is, by the same token, smaller than S'. Because S is nearer than S', it exerts a stronger pull on P; but because it is smaller, it exerts a weaker pull.

The question is to decide how these two aspects balance and Newton showed that, on the assumption that the force of attraction obeys an inverse square law, they exactly cancel out. So P is pulled neither towards S nor S'. By regarding the sphere as made up of infinitely many of these little double cones, you can see that P is pulled in no direction at all: there is no net gravitational attraction inside a spherical shell.

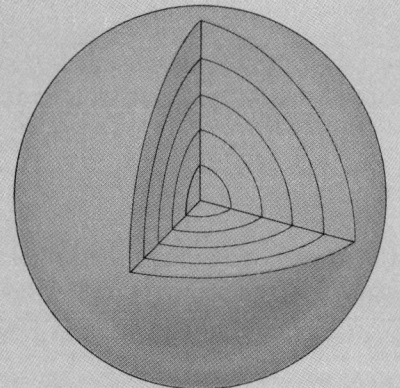

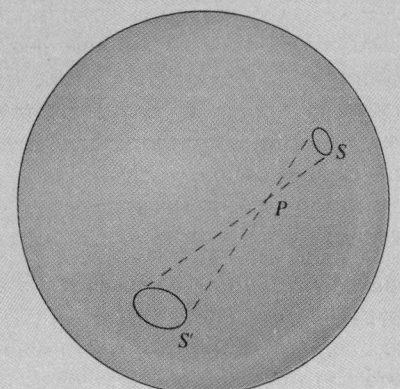

Book II   The Motion of Bodies (in resisting mediums)

This book consists of a discussion of the motion of bodies subject to various forms of air resistance, of hydrostatics and the density and pressure of liquids, pendulums oscillating in resisting mediums, liquids emptying through holes and flowing through canals, and of the transmission of motion through a fluid. These are important subjects, but we shall need to note only that at the end of the book Newton looked at Descartes' theory of motion in vortices and concluded: 'Hence it is manifest that the planets are not carried round in corporeal vortices'. Book II ends with these words:

> the hypothesis of vortices is utterly irreconcilable with astronomical phenomena, and rather serves to perplex than explain the heavenly motions. How these motions are performed in free spaces without vortices, may be understood by the first Book; and I shall now more fully treat of it in the following Book.

SB 12.B9

This is a central accomplishment of the *Principia*. The motion of the planets is shown to be explicable in terms of an intuitively implausible hypothesis (attraction at a distance by a force) which, however, is mathematically derived from, and consistent with, observation. An intuitively plausible physical hypothesis (the vortex theory of planetary motion) is shown to be false on purely mathematical grounds. This is a sizable conjunction of intellectual events, and was to generate a fierce controversy. But, first, how did Newton 'more fully treat of it'?

Book III   The System of the World (in mathematical treatment)

This contains Newton's demonstration that the theory of an inverse square law for gravity acting as a force between bodies can:

(i)   explain how the orbits of planets around the sun, and of satellites around their parent planets, are consistent with Kepler's third law;

(ii)   predict that the planets have elliptical orbits, one focus of which is at the centre of the sun, and move at a rate consistent with Kepler's second law;

(iii)   explain why pendulums of the same length beat with different times at different latitudes, and so enable the shape of the Earth to be determined (Newton argued that the Earth was not spherical, but was flatter at the poles);

(iv)   begin to predict the motion of the moon (an extremely difficult problem and one Newton did not completely resolve);

(v)   account for the motion of comets (specifically, the great comet of 1680).

Newton's work left some questions obscure: why, for instance, the planets all lie in roughly the same plane, and orbit the sun in the same direction. However, because of the retrograde motion of certain comets, this uniformity was no longer complete and therefore failure to account for it was not a major criticism of his theory. Indeed, major changes in scientific conceptual models often involve ignoring questions previously thought important and relevant. Thus just as a question which Kepler thought very significant—why there are only six planets—played no part in Descartes' vortex theory, so some Cartesian questions were of no particular concern (and certainly were unanswerable) in the Newtonian system.

This brief summary of the *Principia* is intended to be helpful, in putting down in one place what you need to know at this stage. It is also meant to overwhelm you, much as one might imagine it overwhelmed its first readers. A lot has been omitted: the mathematical technicalities, all the physics including his discussions of the transmission of sound and all the comparisons between theory and observation that Newton conducted so thoroughly. It would be hard to overestimate the difficulty and originality of the work. It was not just another contribution to the debate about natural philosophy. It was intended to change the grounds of that debate, and so, eventually, it did.

In due course, we shall discuss the reception of *Principia*. But the rest of this section stays with the work itself, looking at it in more detail. Book II has been rather brusquely summarised, for reasons which have to do with its subject matter. Books I and III form an oddly contrasting pair.

THE

CONTENTS.

THAT the matter of the Heavens is fluid   Page 1
The principle of circular motion in free spaces   4
The effects of centripetal forces   5
The certainty of the argument   7
What follows from the supposed diurnal motion of the Stars   8
The incongruous consequences of this supposition   9
That there is a centripetal force really directed to the center of every planet   10
That those centripetal forces decrease in the duplicate proportion of the distances from the center of every planet   12
That the superior planets are revolved about the Sun, and by radii drawn to the Sun, describe area's proportional to the times   15

Figure 6 Newton initially drafted Book III of *Principia* as a more popular, less mathematical account, which was published in 1728 as *A Treatise of the System of the World*. (Here 'World' means 'Universe'.) This is from the second (1731) edition

Question 2   Which book would attract you most if you were a practising astronomer? Which one is most striking to a mathematician, and on what grounds?

Comment ————————————————————————————

The third Book is the one for the astronomer: not only does it account for planetary orbits, but it presents a theory of the motions of the moon and of comets. These theories are tied exceptionally closely to observations. On the other hand, the mathematician would like Book I. The very general account of motion, as well as the specific analyses of motion under various kinds of centripetal force, and the many results about conic sections, are all impressive, as is the dramatic reduction of solid bodies to points for purposes of calculation. The inverse square law of gravity (which is what makes possible this dramatic reduction) is here derived from Kepler's laws as a theorem and not made as an assumption.   ■

We shall see later that it is the existence of two such approaches in one work which accounts in large part for the varied ways in which the *Principia* was received. But let us now look at one or two passages in more detail, to see how these different approaches actually appear.

After the introductory definitions, axioms (the laws of motion) and corollaries— some further implications of the axioms—Book I itself leads off with a series of lemmas about the first and last ratios of quantities. The first three of these lemmas form **SB** 12.B3. Have a look at these now and see if you recognise the issues of concern to Newton here.

These lemmas express the geometrical calculus ideas which, as you saw in the previous unit (*Unit 10*, Section 2), Newton first published in *Principia*. They express in a subtle and plausible form the intuitive idea that you get very close to a curvilinear area by just covering it in narrow rectangles and shrinking their widths more and more. In fact, Newton goes further than this, and says that under these conditions the two areas concerned (the curvilinear area, and its covering by narrow rectangles) *become ultimately equal*. Lemma 1 defines this important notion as what happens when quantities or their ratios 'approach nearer to each other than by any given difference', thus providing a perfectly workable limit concept, in effect, which geometrical arguments can be based upon with no recourse to infinitesimal considerations.

The strength and utility of the concept of *ultimate equality* is soon made manifest, in the very first theorem of *Principia*. This is the important result which generalises Kepler's second (equi-area) law. (**SB** 12.B5, 'Proposition 1, Theorem 1'.) The proof is so remarkably simple using the notion of ultimate equality, and the result is such an important and elegant one, that it is well worth taking as our example of a Newtonian Theorem to study in depth here. We shall work through this together, you reading **SB** 12.B5 while we give a commentary on cassette. Please *switch on the cassette now*.

Newton's Propositions, numbered consecutively, consist of 'Theorems' *or* 'Problems', numbered independently.

Let us turn now to the relationship between Books I and III of the *Principia*. Newton addressed this point in his preface to the whole work.

Question 3   Read the first few lines of **SB** 12.B1(a). How did Newton see Books I and III fitting together?

Comment ————————————————————————————

Book I (and Book II for that matter) describes how to pass from visible phenomena to invisible forces by means of mathematics. Book III is 'an example of this', in that the mathematical propositions of the first Books are there made to 'derive … the forces of gravity', from which forces are deduced 'the motions of the planets, the comets, the moon and the sea'.   ■

It is clear from this preface why Book III is so important. It is only there that the force of gravity is shown to give a really accurate account of the motion of the planets. What might a contemporary reader have found most difficult to understand in Book III?

One non-trivial difficulty would be the technical business of dealing with astronomical data, but suppose we consider a competent practitioner. Then the real difficulty would certainly be the kind of explanation Newton proposed. A *physically real attractive force* reaching out over literally astronomical distances was a colossal, even shocking, novelty in 1687, and is truly still a bizarre concept when you think about it. Leibniz and Huygens could never accept it. Huygens wrote to Leibniz (18 November 1690), 'I am by no means satisfied [by] ... his Principle of Attraction, which to me seems absurd'.

Quoted in Cohen, *The Newtonian Revolution*, p. 80.

Huygens wished to retain Cartesian vortices, not because he could use them to describe orbits to a high degree of quantitative accuracy (he never could, and Newton had shown it was impossible), but because he preferred the conception of dynamics that that theory offered. For him, the *Principia* was elegant geometry, but it was not physics. It is this puzzle—that Newton's most gifted immediate readers were not persuaded, but that their successors were—which we shall now discuss, following the treatment of the Newtonian scholar I. B. Cohen.

Cohen has argued (in *The Newtonian Revolution*) that the answer to this puzzle lies in the bold extension of scientific reasoning that the *Principia* contains. The whole structure of the work is designed to persuade its readers of the effects of this mysterious force of gravity, not in virtue of its essence (whatever that might be) but by a logical mathematical argument. Books I and II are hypothetical. *Suppose*, said Newton, there is such a centripetal force; *then* such-and-such conclusions follow. So far, so mathematical—the supposition is a mathematical idea being entertained for its own sake. But why suppose such a mysterious force in Nature?

It will help to think for a minute how one might come to believe in the existence of anything that one cannot experience directly. One way is the discovery that the idea (that this thing exists) makes sense of a number of disparate phenomena while not simultaneously making nonsense of something else. In the present case, taking as the disparate phenomena the sundry elliptical orbits of planets and satellites as they are described by Kepler's laws, and as the idea that there may be some kind of centripetal force at work, then the shapes of the orbits, and the speeds with which they are traversed, turn out to be consequences of the idea under investigation. In fact, all the orbits turn out to be consistent with the more precise idea that the force must obey an inverse square law. So the idea brings together the astronomical data encapsulated in Kepler's laws by showing that the laws are *mathematical* consequences of supposing there to be an inverse square law of force in operation. The judgement that this hypothesis is not meanwhile making nonsense of something else is rather more subjective. On the hand, the only place where Newton was forced into difficulties in merging theory and observation was the motion of the moon, but there are good reasons (discussed below) for why this does not imperil his overall argument. On the other hand, a reader might feel that the very idea of a force acting at a distance was absurd and could offer no explanation of anything—such an objection might be categorised as 'metaphysical' in the best sense of the word.

On this analysis, the *Principia* aimed to convince its readers of the truth of a physical hypothesis by arguing that it helped to make mathematical sense of certain natural phenomena. Without appreciating this fact, one was naturally liable to read the book as an exercise in mathematics, not in physics. And readers looking for a physical explanation (that is, one which invokes an intuitively plausible physical or mechanical process) would not find one. So the differing receptions of the *Principia* might derive from the differing expectations of its readers about what it had to offer.

However, the reader who could follow Newton's line of thought through all three books would see presented a marvellously well-developed dynamics, truly a System of the World. This, as Newton showed, a vortex theory could never be. Cohen calls the Newtonian style 'a special blend of imaginative reasoning plus the use of mathematical techniques applied to empirical data', and he documents the steady rise in complexity of the objects the *Principia* discusses: a point and a central force; two points mutually attracting each other; two solid bodies; three solid bodies; and the Earth, moon, and sun. By the end, it is the real world that is being described, and not a simplified model of it, or so Cohen argues Newton's view to have been.

Cohen's point about increasing complexity is well taken. A wealth of natural observations were shown to fit a simple mathematical law to a surprising degree of accuracy. Newton's discussion of the motion of the moon was very sophisticated. But here it is the role of mathematics which deserves special attention from us. Three aspects are noteworthy:

(i) technical dexterity (formidable);

(ii) mathematical novelty (or, how the calculus nearly appears);

(iii) hypothetical reasoning (the crucial point).

'Hypotheses non Fingo'—'I frame no hypotheses'—has become Newton's best known saying (**SB** 12.B13). But nothing could be more open to misunderstanding, for the *Principia* is full of hypotheses. All Newton meant was that he framed (i.e. made) no hypotheses about how gravity had its effects. He did not like uncontrolled speculation about mechanisms intended to cause changes, such as he found in Descartes' physics. But he did believe strongly in testing speculations by drawing out their consequences, and the *Principia* is full of arguments of the form

'If X, then Y'

where X is a speculation, a law say, Y a conclusion, such as an orbit, and the deduction is firmly mathematical. To us, but not to Newton, such reasoning is hypothetical; to Newton it was framing hypotheses to slip in a causal explanation without testing it. With this distinction in mind, Newton's *Principia* can be called a showcase of mathematical hypothesis testing.

As such it was novel, and accordingly difficult to read. Books I and II (400 pages out of the 547) are mathematical. They show how more or less any plausible theory of motion can be tested mathematically by being made to predict orbits, via his theorem on equal areas in equal times. What emerges is that no *simple* laws of motion other than motion under an inverse square law of attraction has any chance of corresponding to Kepler's laws. Having thus driven rival theories from the field, it is only in Book III that Newton turns to physics, and shows that inferences on the assumption of gravity do indeed correspond exceptionally well to empirical evidence. Newton was clear that you had to accept gravity *because*

(i) there were no other theories which were mathematically sound, and

(ii) gravity yielded descriptions well in accordance with observations,

even though

(iii) there was no physical explanation available for how gravity operated.

This is quite different from how the vortex theory operated. There a simple mechanical explanatory model (collision) was intended to give a rough fit with observational data, and the model retained its appeal even though no-one could make it yield a good fit. The idea that a range of mathematical theories be put on offer, and the best fit selected, is a novelty one sees first with Newton. If it resembles anything, it resembles Kepler's choice of the ellipse as a curve for fitting planetary orbits, but raised to a higher level of sophistication.

Newton distinguished carefully between mathematical principles and their physical application. What is described in Books I and II, including an attractive force obeying an inverse square law, is mathematical and need not be to do with the real world. His attitude to the nature of gravity was to be phrased very carefully, not in the first edition, it is true, but in the second and third editions (of 1713 and 1726), which were the ones to be read most diligently by subsequent generations. We may presume that the lack of much comment by him on the subject initially was noticed by his very first readers, and that what he was later to say about gravity was the product of much attentive thought.

Question 4 Read **SB** 12.B13. On the basis of what Newton says there, what do you think he came to believe about gravity? Did he consider it really existed, or that it was but a useful descriptive device?

Comment —————————————————————————————————————

He argued that gravity explained the motion of the planets, although how it acts ('the cause of this power') he had been unable to discover, nor would he speculate. But he was adamant that gravity really existed and obeyed the laws he had set down. It was not a mathematical artifice, although mathematics was needed to discover it, but a physically-existing thing. ∎

Before pursuing, in the next section, how *Principia* was received, we conclude this one by briefly considering the mysterious case of the non-appearing calculus. It was Newton himself who started the story that the contents of *Principia* were discovered through using the calculus, and then re-written in more traditional geometrical language:

> By the help of the new *Analysis* Mr Newton found out most of the propositions in his *Principia Philosophiae*: but because the Ancients for making things certain admitted nothing into Geometry before it was demonstrated synthetically, he demonstrated the Propositions synthetically, that the Systeme of the Heavens might be founded upon good Geometry. And this makes it now difficult for unskilful men to see the Analysis by which those Propositions were found out.

Cited in Westfall, *Never at Rest*, pp. 423–4.

This is an interesting example of how primary historical testimony may be misleading if taken at face value. It appears that this version of history is what Newton wished to be believed, for reasons associated with the sad dispute of his later years concerning priority in inventing the calculus. But as the editor of Newton's papers, D. T. Whiteside, has written,

> Nowhere, let me repeat, are there to be found extant autograph manuscripts of Newton's, preceding the *Principia* in time, which could conceivably buttress the conjecture that he first worked the proofs in that book by fluxions before remoulding them in traditional geometric form.

D. T. Whiteside, 'The mathematical principles underlying Newton's *Principia Mathematica*', *Journal of the History of Astronomy* **1** (1970), pp. 125–6.

What is true is that the same gifted mind that could produce a theory of fluxions produced the *Principia*—and in the same way, by dealing geometrically with finite quantities and approximation arguments. Not that Newton was a particularly well-read geometer; in this as in so much else he was very much his own man. And it is true that Newton could derive some propositions in the *Principia* more easily with his calculus than without it. In a famous scholium he indicated that he had, since at least 1671, possessed a general method for finding tangents to curved lines and 'to the resolving other abstruser kinds of problems about the crookedness, areas, lengths, centres of gravity of curves &c'. (Lemma II of Book II, to which this note was attached, amounts to differentiating $x^{m/n}$ to get $\frac{m}{n}x^{(m-n)/n}$, and is indicative of the high level of generality to which Newton was working.) Although the *Principia* is not a calculus book, the questions it discusses are appropriate for the calculus. Newton's successors, lacking his geometrical brilliance, put into their understanding of his work the calculus-type arguments he had eschewed. Mathematical physics is written in terms of the calculus not because of Newton's example, but because without it no-one else could easily follow him.

A *scholium* is an explanatory or historical remark, of which Newton inserted several into the *Principia*.

11.3 THE IMPACT OF THE *PRINCIPIA*

The appearance of the *Principia* caused a sensation. R. S. Westfall has noted some contemporary reactions. David Gregory wrote from Edinburgh that Newton had

> been at the pains to teach the world that which I never expected any man should have knowne.

Edmond Halley said that

> so many and so Valuable *Philosophical Truths*, as are herein discovered and put past Dispute, were never yet owing to the Capacity and Industry of any one Man.

Westfall, *Never at Rest*, p. 470.

John Aubrey, in vain pursuit of the priority claim on behalf of his friend Hooke, spoke of

> the greatest Discovery in Nature that ever was since the World's Creation. It never was so much as hinted by any man before.

O. Lawson Dick (ed.), *Aubrey's Brief Lives* (Peregrine Books, 1962) p. 245.

And when the Marquis de l'Hôpital was shown the *Principia* by Dr Arbuthnot,

> he cried out with admiration Good god what a fund of knowledge there is in that book? he then asked the D^r every particular about Sr I. even to the colour of his hair said does he eat & drink & sleep. is he like other men?

Westfall, *Never at Rest*, p. 473.

It had a dramatic impact on the philosopher John Locke, then a political exile in Europe. Locke found the mathematics beyond him, but was assured by Huygens that he could take it on trust, so he went directly to the physics, which greatly impressed him. Quite generally, the scope of the work, coupled with its obvious difficulty, gave it an impressive reputation far outside the limited circle of mathematicians competent to read it.

However, if its publication was a major event, its subsequent readings are not so easy to characterise briefly. Mathematically, the *Principia* is fiercely difficult, and readers lacking Newton's brilliance at geometry had to resort to (and thus to have learnt) the calculus in order to master the work. In this way, it marks a turning point in the development of mathematics. However, to practitioners of science the book was a conundrum, the problem being the nature of the attractive *force of gravity*. We have already noted Huygens' refusal to accept such a force. Although he wrote in 1688 'Vortices destroyed by Newton', he could only try to replace them with other vortices of a different, non-Cartesian, kind about which his thoughts remained (understandably) quite vague. Leibniz, basing himself not on the *Principia* but only on a review of it published in *Acta Eruditorum* (1688), attempted a more detailed defence of vortices, which was marred by a number of major errors. Even when he tried again, Huygens had to point out that his revised vortex theory could not accommodate Kepler's third law (the $\frac{3}{2}$ power law). What this prodigious attempt by the two leading Continental mathematicians tells us is how firmly they resisted the attractive force of gravity. To them it was an 'absurd' idea (Huygens' word), incapable of *explaining* anything.

Twenty-five years later, when bad feelings between Newton and Leibniz erupted into an open feud, they fought over the nature of gravity as well as over the invention of the calculus. Leibniz continued to argue that it was not enough to say, as Newton did, that the cause of the properties of gravity had not been discovered. This was, in his view, an evasion designed to avoid confronting the implausible mechanism by which gravity operates. Newton thought this was mere playing with words, and when Leibniz criticised the *Principia* for not explaining gravity, a force Leibniz found implausible and obscure ('occult' was his word), Newton in 1715 drafted a reply complaining that,

> He denys conclusions without telling me the fault of the premisses ... His arguments against me are founded upon metaphysical & precarious hypotheses & therefore do not affect me: for I meddle only with experimental

Philosophy ... He changes the signification of the words Miracles and Occult qualities that he may use them in railing at universal gravity.

Cited in Cohen, *The Newtonian Revolution*, pp. 61–2.

The issue was never to be completely resolved. Action at a distance is still thought to require an explanation (Einstein's general theory of relativity being the most widely accepted candidate), but collision mechanisms rather fell out of favour. To understand why, it is not enough to look at their poor showing even in the hands of Leibniz and Huygens. It is also necessary to look at the French school of mathematicians during this time.

The French were neutral in the Anglo-Hannoverian squabble over who invented the calculus, inclining to one side or the other as they saw fit. On matters of physics they were firm Cartesians; so it is their change that is the most interesting. In 1688, Pierre-Sylvain Régis' review of the *Principia* (in the *Journal des Scavans*) set a rather hostile tone:

one cannot regard these demonstrations otherwise than as only mechanical; indeed, the author recognises himself that he has not considered their Principles as a Physicist, but as a mere Mathematician.

Cited in Cohen, *The Newtonian Revolution*, pp. 16–7.

Question 5 How accurate do you feel is this comment about *Principia*?

Comment ───────────────────────────────────────
That this is a mis-reading should be clear. Régis has correctly seen that Books I and II are hypothetical (in our sense) and mathematical, but he has failed to see how those mathematical results are applied to the natural world in Book III. ■

Let us see how Newton's use of mathematics came to be better appreciated. The central figure is the Oratorian Father Nicolas Malebranche (1638–1715) who came to mathematics via his discovery of Descartes' philosophy in 1664. A well-read mathematician, who considered mathematics 'the foremost and fundamental discipline of all the human sciences', he gathered a notable group of people around him: Jean Prestet, Charles-René Reyneau, Pierre Rémond de Monmort, the Marquis de l'Hôpital, and, peripherally, Pierre Varignon. Of these, it was de l'Hôpital who brought the Leibnizian calculus to France, via his association with Johann Bernoulli, and Varignon was to be the one who dealt most ably with the motion of a body under various forms of central force.

It was Malebranche himself who made the first break with Cartesian ideas. After writing an *Elémens des mathématiques* with Jean Prestet in 1675, he subsequently disavowed its denial, which had followed Descartes, that mathematics could deal with the infinite. Malebranche thereby opened up a willingness to employ calculus-type reasoning in many problems. His emphasis on mathematics as 'the most exact and unimpeachable form of knowledge' seems to have inclined Malebranche and his school to appreciate Newton's radical abandonment of hypothetical mechanisms in physics in favour of mathematical deduction. They quickly realised that Newton's views had to be taken seriously, and were in the forefront of attempts to elucidate them by means of Leibniz' calculus. Under Malebranche's influence Varignon moved towards considering forces in a Newtonian fashion, while never quite freeing himself of Cartesian vortices. His position involved a species of agnosticism about underlying physical causes. Slowly but steadily this school came to appreciate the logical structure of the *Principia*, while attempting to retain something of the vortex theory.

Quoted in H. Guerlac, *Newton on the Continent* (Cornell University Press, 1981) p. 59.

The process was continued by the next generation. Jean-Jacques Dortous de Mairan and Joseph Privat de Molières, both disciples of Malebranche, made ingenious attempts at reconciling the incompatible theories of Descartes and Newton—from which emerged a lively sense of the power of mathematics, and an incipient realisation of the validity of the Newtonian style of reasoning. This produced an eclectic and unstable liaison of Cartesian, Malebranchist and Newtonian ideas. It could not last, and the generation of the 1730s was to sweep it away, but Cartesian physics had been effectively undermined. As Emilie du Châtelet wrote of Cartesianism in 1738: 'It is a house collapsing into ruins, propped up on every side ... I think it would be prudent to leave'.

T. Besterman (ed.), *Les lettres de la Marquise du Châtelet*, vol. I (Institut et musée Voltaire, 1958) p. 261.

Figure 7 Emilie de Breteuil, Marquise du Châtelet (1706–1749)

Figure 8 Voltaire (1694–1778)

For the remainder of this section we shall discuss later responses to the *Principia*. Although the great success of Newton's *Principia* was its discussion of planetary motion, in fact this success was slightly flawed by the imperfect account of the motion of the moon. The treatment of many other topics was also somewhat contrived, although an improvement on contemporary practice. For example, the calculation of the shape of the Earth (which Newton took to be treatable as a fluid) depended on implausible and *ad hoc* assumptions about the nature of fluid pressure. Rather more surprisingly, Newton's laws of motion did not immediately present themselves in a form appropriate for mathematics, a point we shall look at in more detail in the last section of this unit.

Two problems, out of many, are worth considering as illustrative of how the *Principia* was received early in the eighteenth century. These are:

(i) the shape of the Earth;

(ii) the motion of the moon.

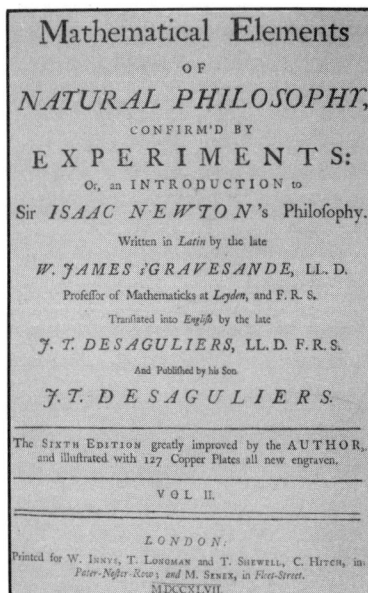

Figure 9 Title page of Willem's Gravesande's *Mathematical elements of natural philosophy confirmed by experiments, or an introduction to Sir Isaac Newton's philosophy* (1720)

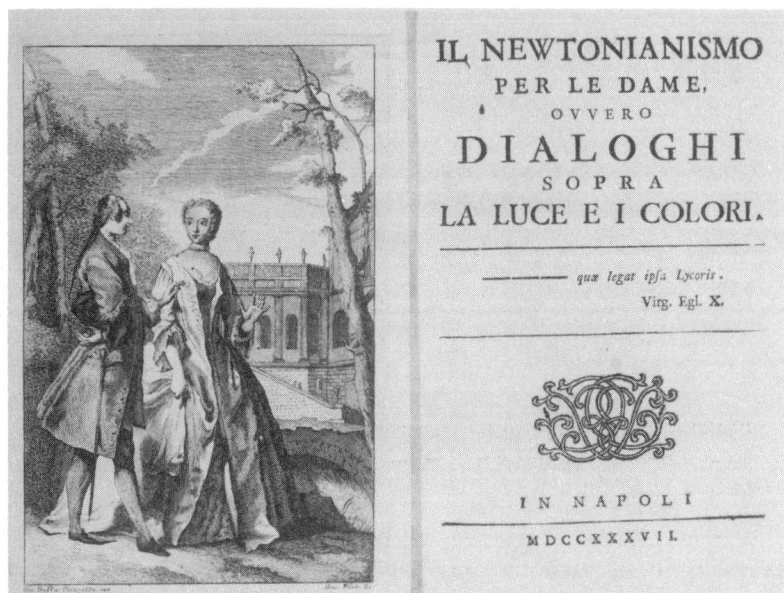

Figure 10 Title page of Francesco Algarotti's *Newtonianism for the ladies* (1737)

Each of these, for different reasons, was a problem for Newton's successors to tackle. Together they will enable us to answer the question: how did Newtonian mechanics become the dominant scientific paradigm of the eighteenth century?

The British took to it very quickly. So did the Dutch school of mathematicians— Willem 'sGravesande wrote a successful book on Newton's ideas in 1720–and so did the Italians. But these groups were smaller and less important than the French one, based in Paris, and the essentially Swiss mathematical community (to which Euler belonged) centred around the widespread Bernoulli family. We shall concentrate chiefly on the French, because it was their conversion to Newtonianism that was to be decisive. Newtonian ideas ultimately became widely accepted. Nature in all its guises, as it was presented in books in the later eighteenth century, became more and more full of attractive or repulsive forces centred on numerous hard bodies—for all the world like miniature solar systems—and the greatest exponent of this point of view by 1800 was the French mathematician, Pierre-Simon de Laplace.

An interesting and vivid light is thrown on Anglo-French relations by one of Voltaire's early essays which described what he saw on his visit to England in 1727, the year of Newton's death.

Question 6   Read **SB** 12.F2. Does Voltaire seem to support in the end, his fellow countryman Descartes, or Newton?

Comment ————————————————————————————

He was certainly struck by the extent to which Newtonians and Cartesians disagreed. While the national chauvinism amused him considerably, he seems to take Newton's side against Descartes. Descartes was praised for being the first to tackle certain problems, but was also said to have been wrong, for example about dynamics and about the nature of matter. The implication is that Newton's views were an improvement. (Incidentally, Descartes' philosophy is again labelled fictional—an 'ingenious novel'—a shrewd reflection on the balance to be struck between thought and experiment.)   ■

Try to find time to look at an extract from the work mentioned by Voltaire, Fontenelle's *Eulogy* of Newton (**SB** 12.F1), for it is an exceptionally thoughtful piece of writing. But let us stay with Voltaire for a moment. The question of the shape of the Earth, which he raised, is far from an idle one.

Question 7   Read Maupertuis' account (**SB** 14.B1). Why would it be useful to know the shape of the Earth, and why would it be helpful to understand the motion of the moon?

Comment ————————————————————————————

In a word: navigation. A pre-requisite for accurate maps is knowing the shape of the Earth; without that knowledge exploration can be perilous. Sailors also require to know their position, and Maupertuis noted that the determination of longitude would seem to require knowing how the moon moves. So both our problems have a practical side.   ■

The shape of the Earth

Because the question of the shape of the Earth was so important, it had been investigated for some time. Voltaire gave a characteristically polemical account of the story, which also explained how the shape could be investigated. Please *read it now*, in **SB** 12.F3.

Question 8   On the basis of this extract, how do you think Newtonians would describe the shape of the Earth, and why? What would they regard as the most likely behaviour of a pendulum at polar latitudes?

Comment ————————————————————————————

To a Newtonian, because a pendulum beats faster in the latitudes of France than it does near the equator, it follows that the Earth is broader at the equator. This is because the greater the force of gravity upon a pendulum the faster it beats, and this force increases as you move towards the centre of the Earth. The most plausible hypothesis would be that the Earth would continue to flatten as one travelled North, so that the poles would be even nearer the centre than France is;

consequently a Newtonian would predict that a pendulum would beat even faster at the poles. ■

Other theories of matter and weight would make different predictions. Huygens' theory did so, for example, as did several other Cartesian ones, so it was possible for observation to discriminate between them. Moreover, one could look at other planets for clues: Newton observed (see **SB** 12.B11) that Jupiter was flatter at the poles. However, later measurements of latitude made in France suggested the opposite result, namely that the polar circumference of the Earth was greater than the equator. Finally, Maupertuis proposed to resolve this question, and he and Clairaut led an expedition to Lapland to measure the size of a degree of latitude in those inhospitable northern conditions, while others performed the same task in Peru, on the equator.

Box 5 Measuring the shape of the Earth

To measure the size of the Earth, assuming it is spherical, you take two observations of the same star at the same time of day, but at two places, one of which is due north of the other. The difference in elevation of the star equals the amount, in degrees of latitude, by which the one place is north of the other. Knowing this distance, you can calculate the size of a degree, and hence the size of the Earth.

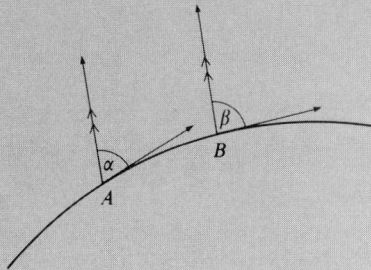

The difference in elevation, $\beta - \alpha$, tells you the difference between the latitudes of B and A

The same calculations, repeated at different latitudes, will only give the same answer if the Earth is a sphere. If it is not, the discrepancy between one set of observations and another, taken between two pairs of points the same distance apart, will enable you to calculate how the change in elevation varies with position, and so enable you to determine the shape of the Earth.

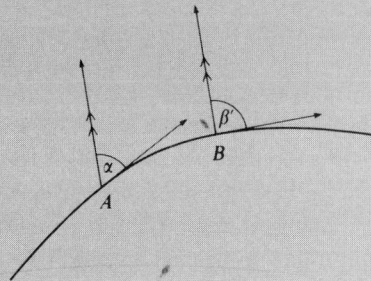

A comparison of β (above) and β' tells you how unspherical the world is near B

The Peruvian expedition took a long time to report, but in 1739, Maupertuis was able to show that Newton was right: the Earth *is* flatter at the poles. Historians have assessed the significance of Maupertuis' work in different ways. A. Rupert Hall regards it as having 'removed the last serious factual obstruction to the universal acceptance of Newtonian mechanics', although he goes on at once to say 'Acceptance, that is, as a basis for further investigations'. The historian Clifford Truesdell thinks it was not so simple, however. Writing of Maupertuis' result, which earned him the nickname of 'the greater flattener', Truesdell says 'While in the popular view the Newtonian system was by this one blow proved correct, the measured eccentricity did not agree with Newton's value, and the geometers were moved to read more critically the passage in which Newton derived his result. They found his argument insecure'. It emerged that although his prediction was qualitatively correct, Newton's *ad hoc* assumptions about fluid pressure were wrong, and gradually attempts were made to tackle this problem too.

A. R. Hall, *The Revolution in Science, 1500–1750* (Longman, 1983) p. 352.

C. Truesdell, 'A program toward rediscovering the rational mechanics of the Age of Reason', *Archive for History of Exact Sciences*, **1** (1960) p. 19.

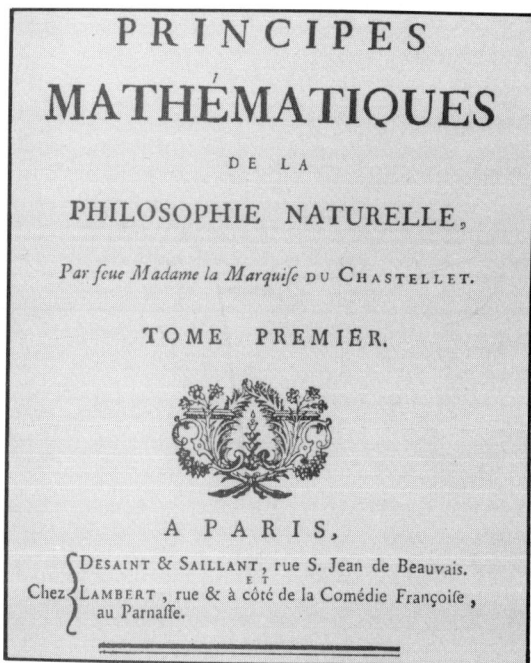

Figure 11 Title page of du Châtelet's translation of Newton's *Principia*, published in 1759, some years after her early death. It remains the only French translation of *Principia*

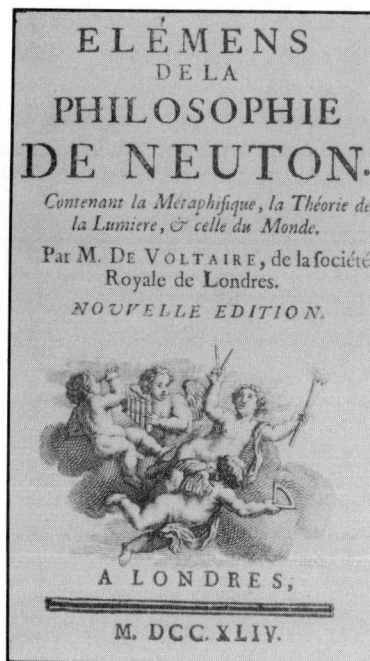

Figure 12 Title page of Voltaire's *Elements of the philosophy of Newton* (1738), edition of 1744

Voltaire's growing enthusiasm for Newtonianism, which can be seen in his increasingly partisan expositions of Newton's ideas, was widely shared. He tells us that he learned what he could understand of the *Principia* from Madame du Châtelet, who, with Clairaut, later translated the *Principia* into French. Indeed, throughout the 1730s there was a fashion for Newton's work which went quite beyond the bounds of those capable of discussing it with any competence. However, as so often in life, the honeymoon did not last. By the end of the 1740s, Newton's theory of gravity was under sustained attack from the three people who best understood it: D'Alembert, Clairaut, and Euler. The problem is our second one.

The motion of the moon

The problem is that the moon is part of a system of three bodies: the Earth, the moon and the sun. While Newton could deal well with two bodies acting on each other by gravity, the *three-body problem*, as it is known, is (strictly speaking) unsolved to this day. That is to say, no-one can yet answer the question: given three arbitrary bodies acting on each other by gravity and released initially with such-and-such velocities, what will be their orbits? We do not know, for example, if the moon will always orbit the Earth, or whether it will move away or one day collide. This mathematical problem is too difficult to solve exactly. On the other hand, one can reach some conclusions if simplifying assumptions are made. If one assumes, for example, that the only effect of the sun is to perturb slightly an otherwise elliptical orbit of the moon around the Earth, one can try to calculate that perturbation exactly. This is what Newton did; it is what everyone has done since. In his case success was less than complete. The moon is an easy object to observe, and the mis-match between mathematical prediction and physical actuality was apparent.

The problem was a technical one. If there were just the moon and the Earth (the two-body problem), the moon's orbit would be an ellipse. The effect of a third body (the sun) is that the whole elliptical orbit moves slowly around (see Figure 13). The question was: *how* slowly? Newton's calculations showed the orbit returning to its original place every eighteen years. But Nature, it appeared, preferred nine. Newton conceded this in later editions of the *Principia*, in a single crisp sentence: 'The apse of the moon is about twice as swift'.

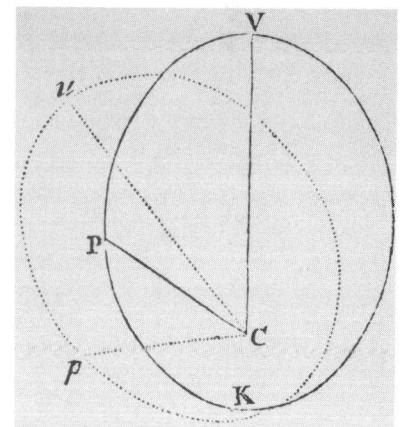

Figure 13 Two positions of an elliptical orbit as it revolves around one focus, *C*; adapted from a diagram in Newton's *Principia*

Principia I, 45, Corollary 2. The *apse* refers to the point on its orbit at which the moon is at its greatest distance from the Earth; the far end of the major axis of the ellipse, in effect (the *V* or *u* of Figure 13).

But while this point may be technical, its implications were not. This failure of Newton's might be, it was thought, the loose thread unravelling his theory. We have seen how unpopular initially his idea of universal gravitation was in some quarters. Newtonian gravity (more precisely, the inverse square law) now came to stand or fall by its ability to describe sufficiently accurately the motion of the moon.

There was an eminently practical reason for caring about the moon: the determination of longitude on ships at sea. Latitude is relatively easy to determine from the sun or the stars, but longitude is not. From 1714 the British Board of Longitude offered a large cash prize for the solution of this problem, which was acute for any sea-faring nation, since the methods then in use produced very large errors (10% or so). Could the heavens be used as a celestial clock? If they could be read accurately the problem of longitude would be solved, whence the demand for an accurate theory of the motion of the moon.

Significant progress on that question had to wait until 1747, when Clairaut proposed to modify the inverse square law by adding on a small inverse fourth power term. That such a proposal could be made shows the uncertain hold the inverse square law had on working mathematicians. His rival d'Alembert came independently to the conclusion that Newton's theory was incorrect, and so did Euler the next year, after he had studied a different 3-body system (the sun, Jupiter, and Saturn), although each man had a different remedy in mind. For a while, Newton's theory of gravity seemed about to fall.

Clairaut suggested replacing an equation like $f = ar^{-2}$ with one of the form $f = ar^{-2} + br^{-4}$.

You will be able to follow what proved an exciting and dramatic debate quite closely if you look at the extracts from the Euler–Clairaut correspondence (SB 14.B2 and B4). When you read the first one, you should know that Euler had just submitted his essay on the motion of Saturn to the Académie des Sciences for consideration in their prize competition. Entries were anonymous, hence the jocular opening remark by Clairaut who was one of the judges.

Question 9   On the basis of **SB** 14.B2(a) and (b), what modifications did Clairaut and Euler propose to make to Newton's inverse square law? Did they 'frame hypotheses' concerning them?

Comment ————————————————————————————
Both proposed to depart from an inverse square law, but whereas Clairaut would have liked to replace it with something like r^{-2} plus a term negligible when r is small, Euler wished to re-introduce vortices. Since only Euler sought to explain gravity with a new mechanism, only he framed a hypothesis. ■

When you look at **SB** 14.B2(c) you will see what Clairaut thought of the re-introduction of vortices, and also what his counter-proposal was.

The reference at the end is to Euler's *Introductio in analysin infinitorum*, then being printed at Lausanne.

Question 10   What did Clairaut now think?

Comment ————————————————————————————
He had withdrawn the suggestion that the modifying term should be an inverse fourth power, because it predicted that objects near the surface of the Earth should be heavier than they are. But he had no confidence in Euler's vortices, which he thought Euler himself had shown to be no help at all. ■

Please *read* **SB** 14.B3 *now* which gives a clear picture of a first-rate scientific mind at work—and it ends dramatically.

Question 11   Summarise in a few words:

(i)   what Clairaut thought made the *Principia* difficult to understand,

(ii)   what Clairaut's views were during 1747 about the inverse square law and

(iii)   what he came to think in 1749.

Comment ————————————————————————————
His complaint is very clear—Newton did not use enough words. One gets a vivid picture of even Clairaut struggling to work his way through the *Principia* in order to understand it rather than merely admire it. No matter, said Clairaut, that Newton concealed his fluxional calculus, for he points out that the calculus is now so familiar that it is easy to repair that omission—of course, Clairaut used the Leibnizian calculus. But (i) for Clairaut it was a matter of regret that Newton did

not explain his principles, especially when one of his predictions turned out to be so obviously wrong. However, Clairaut reflected, much else is right, so (ii) it must be subtle change that is required in the law of gravity, and that is what he set out to discover. But (iii) then comes the surprise. On 17 May 1749 he simply announced that, by taking a new point of view, the problem disappeared, and the inverse square law could give the correct prediction for the apse line of the moon. ∎

So quite suddenly Clairaut retracted his idea altogether. D'Alembert, who had waited for Clairaut to declare his hand, thereupon retracted too. What, then, was the new viewpoint which was so powerful? It was an improved mathematical analysis of the problem, specifically a new approach to the differential equations which were taken to describe the motion of the moon—so it was another victory for the calculus.

Clairaut now proposed instead that the error lay in the poor way in which exact but unsolvable equations for the motion of the moon had been reduced to inexact, approximate, but solvable equations. Euler did not at first agree, and proposed that the St. Petersburg Academy establish a prize competition on the subject. It did, and when Euler satisfied himself that Clairaut was right, he saw to it that Clairaut received the prize. He then published his own theory of the motion of the moon in 1753.

The practical consequences were immediate. Tobias Mayer, an astronomer at Göttingen, calculated a set of lunar tables based on Euler's theory in 1755, and eventually, in 1765, the British Board of Longitude rewarded his widow with £3 000 and paid Euler £500 for his contributions. At the same time, they paid the English inventor J. Harrison £20 000 for the usable solution. As those of you who have seen it in the museum at Greenwich may recall with pleasure, that solution is an accurate, spring-driven, pocket watch.

We shall not go into the details of Clairaut's method, which even Euler found difficult, as you can see from **SB 14.B4**. Rather, we shall conclude by observing that Newtonianism became accepted very much as Newton had presented it, in the following senses:

(i)   the theory of the solar system was highly mathematical;

(ii) the predictions it made rested on a highly theoretical analysis;

(iii) its theoretical presuppositions seemed inevitable if its conclusions were accepted, in particular, the mysterious force of gravity was accepted as a real, existing thing.

The role of mathematics in science was now much greater than it had been, for it became the glue holding the vast new edifice together. For us, as historians of mathematics, that is perhaps the most important consequence of the growing acceptance of Newtonianism in the period 1687–1749. We shall therefore look at this role in more detail in the next section.

Figure 14 This watch by John Harrison, 1761, is about 6″ (c.15 cm) across

11.4 RATIONAL MECHANICS

What role did mathematics play in the eighteenth century in providing causal accounts of natural phenomena? We consider first what kind of phenomena were involved. Was there a strong desire for mathematical input to technological developments? On the whole, it does not appear that there was. This may seem rather surprising. So accustomed have we become to the utility of science and its crucial role in modern production, to debates about the need for research to produce new techniques and commodities, and to the vast laboratories and proving grounds of technology and the ubiquity of its products, that we naturally transfer this complex back to the past. Surely, we say to ourselves, it was like that then, too, albeit on a smaller scale. The surprising truth is that it was not. There seems to have been almost *no* pressure on scientists to produce 'useful' results. While it is always difficult to prove a negative—even a weak, modified one like this—a look at the kind of useful work that was done will perhaps make clear how much the eighteenth century differs from our own in this respect.

The ever-prolific Euler did involve himself in applicable work. He designed a turbine, for example, in 1754—but it was never built. And although he did important work on hydrodynamics, that was quite independent of the design of canals on an industrial scale, and he went no further than to design an aqueduct to amuse Frederick the Great. His book *Naval Science* (*Scientia Navalis*, 1749) gave his customarily lucid account of the topic at hand, in this case buoyancy and fluid pressure. Revised and abridged in 1775, the new version was translated into English in 1776, which suggests that it was regarded as important by those who ran the world's foremost naval power. He also investigated one other subject that, then as now, is of interest to governments: gunnery. The problem Euler considered was this. Air cannot rush in behind a projectile faster than it can enter a vacuum, but a projectile can move faster than that critical speed. When it does a partial vacuum opens up behind it, and the problem is to understand how the projectile will then move.

The English engineer Benjamin Robins, who worked for the East India Company, wrote a practical tract *New Principles of Gunnery* in 1742. It came to the attention of Frederick the Great, whose entire political life was spent seeking to enlarge his small country by waging war on his larger neighbours. Frederick was therefore much interested in military innovations, so he asked Euler to take up Robins' work. Euler responded in 1745 with a much extended version of the tract (which in its turn was translated back into English in 1777). It is reassuring to note that Condorcet in his *Eloge* of Euler found that this work advanced nothing except the science of calculation. The topic remained of interest, however, for in 1766 a younger member of Frederick's Academy of Science, Johann Heinrich Lambert, took it up, and sought to give a better theoretical account of the motion of a projectile. His investigation of gunnery, like all his work, is a thorough re-working of basic principles. Whether Lambert devoted much attention to the ethical implications of this piece of research is difficult to say, not least because the tradition is quite an old one whereby the theoretical side of military science is dissociated from considerations of its practical consequences.

It is hard to see these studies as central to the lives of either man, and there are no examples of other eminent mathematicians busying themselves any more than did Euler and Lambert with questions of applied science. So for reasons like this we can analyse the development of applied mathematics in the late eighteenth century without involving ourselves any further in its utilitarian and technological aspects.

What then, of the application of mathematics to natural phenomena such as heat diffusion, electricity, and magnetism? It does not appear there was a strong drive in this direction either. Despite some early work by Lambert and others, the first worthwhile mathematical accounts of heat conduction were not to be given until the work of Joseph Fourier at the start of the nineteenth century.

There was an older tradition of investigating magnetism and electricity. The case of electricity is typical. Once easily repeatable phenomena were discovered in the 1740s, electricity became the leading branch of experimental physics, though it seems that many of these experiments were more recreational than instructive. One teacher, A. G. Kaestner, even gave up teaching via experiment because his students only 'wished to see physics, not to learn anything about it'. But the students of electricity who emerged late in the century were not mathematically inclined and mathematics was of no particular advantage in the study of their subject. The problem was that electricity proved difficult to quantify; indeed, until the 1790s the study of electricity was regarded as an experimental science and not as a mathematical one.

Throughout the eighteenth century the core of the mathematical sciences remained what it had been for so long: the quadrivium studies of astronomy and music together with optics, statics and—since the Middle Ages—the study of motion. These, adjoined to the fundamental studies of arithmetic and geometry, had stayed together as a unified complex, in the sense that a mathematically-educated person would be competent in all these things. The longevity of this cluster of studies is remarkable, as the historian Thomas Kuhn has noted:

24

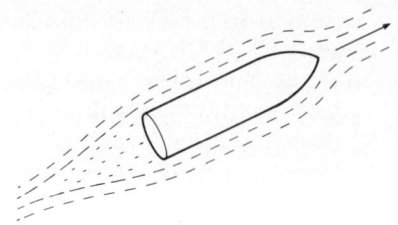

Figure 15 As the projectile rushes forward a partial vacuum opens up behind it; this phenomenon is called *cavitation*. The difference in air pressure front and back increases the resistance from the air to the object's motion

King Frederick II of Prussia (1712–1786)

SB 14.C4

Lambert (1728–1777) is an interesting figure, a self-taught Swiss polymath whom we shall meet in *Unit 13* investigating non-Euclidean geometry.

Abraham Gotthelf Kaestner (1719–1801) was a friend of Lambert, and later Gauss' first professor at Göttingen.

Quoted in J. L. Heilbron, *Elements of Early Modern Physics* (University of California Press, 1982) p. 8.

The classical sciences continued from the Renaissance onward to constitute a closely knit set. Copernicus specified the audience competent to judge his astronomical classic with the words, 'Mathematics is written for mathematicians'. Galileo, Kepler, Descartes, and Newton are only a few of the many seventeenth-century figures who moved easily and often consequentially from mathematics to astronomy, to harmonics, to statics, to optics, and to the study of motion. With the partial exception of harmonics, furthermore, the close ties between these relatively mathematical fields endured with little change into the early nineteenth century, long after the classical sciences had ceased to be the only parts of physical science subject to continuing intense scrutiny. The scientific subjects to which an Euler, Laplace, or Gauss principally contributed are almost identical with those illuminated earlier by Newton and Kepler. Very nearly the same list would encompass the work of Euclid, Archimedes, and Ptolemy as well. Like their ancient predecessors, furthermore, the men who practiced these classical sciences in the seventeenth and eighteenth centuries had, with a few notable exceptions, little of consequence to do with experimentation and refined observation even though, after about 1650, such methods were for the first time intensively employed to study another set of topics later firmly associated with parts of the classical cluster.

T. S. Kuhn, 'Mathematical versus Experimental Traditions in the Development of Physical Science' in *The Essential Tension* (University of Chicago Press, 1977) pp. 39–40.

The centrality of mathematics in a tradition of such antiquity, reinforced as it was by the success of Newtonian physics, helps us understand the unexpected difference between eighteenth-century scientific practice and that of own day. The point is not that mathematicians were trained in a wide variety of 'applied' disciplines, but that such a breadth of activities *constituted* mathematics. To obtain a more precise grasp of the role of mathematics, it will be helpful to consider a typically vigorous description of the situation by Clifford Truesdell. In a celebrated article written in 1960 as part of a campaign to generate historical studies of the exact sciences, he said: '[anyone] who is trained in physics today will ask, what were the fundamental experiments upon which 'classical' mechanics was founded?' Truesdell's answer, founded on his extensive reading of the eighteenth-century literature, was stark and simple: 'I have never been able to find any'. In what will be the leit-motif for this section, Truesdell went on:

> What was, then, the method? Rational mechanics was a science of *experience*, but no more than geometry was it *experimental* ... While some great mechanical experiments were done in the Age of Reason ... [and there] were also large, cooperative projects ... the effect of all this expense on what we now consider the achievement of the period was nil. The method used in the great researches was entirely mathematical, but the result was not what would now be called pure mathematics. *Experience* was the guide; *experience*, physical experience and the experience of accumulated previous theory. If we were to seek a word for what was done, it would not be physics and it would not be pure mathematics; least of all would it be applied mathematics: it would be *rational mechanics*.

Truesdell, 'A program toward rediscovering the rational mechanics of the Age of Reason', p. 36.

Although there are several things to tease out of this paragraph, for example, the distinction between experience and experiment, Truesdell's broad brush picture is surprisingly convincing. The arbiter in scientific debate throughout the eighteenth century was generally the theorist, seldom the experimenter. The balance only shifted in the nineteenth century. And the evidence the theorists would adduce was experience, not experiment.

Question 12 Consider the subjects discussed earlier in this unit. Why might *experience* predominate over *experiment* in the investigation of the matters raised by Newton?

Comment —————————————————————————————
Gravity does not lend itself to experimental study, only to passive observation. The motion of the moon and the shape of the Earth are hardly to be altered in the laboratory, even today, but can be dealt with by a combination of measurement and mathematics. ∎

What of Truesdell's other distinctions? When he distinguishes between physics and pure mathematics he is separating out an approach to the study of nature which is driven by experiment, from a formal, logical system of deductive reasoning. This distinction was widely observed in the eighteenth century as that between the classical, theoretical sciences and the new, experimental ones. His rejection of the term 'applied mathematics' derives partly from a wish not to import a twentieth-century term back to the eighteenth, but also because applied mathematics is a subject in which mathematics plays a subservient role. Mathematics is applied, usually as a technique to harmonise discoveries already made. It might make the theory more elegant, but it is not creative. The term 'rational mechanics' is meant to direct our attention to the subject matter— mechanics (Truesdell did not discuss optics, alone of Kuhn's classical sciences) and to the manner in which it was discussed: rationally, theoretically, and in a significant way, mathematically.

The size of the rational mechanical enterprise is itself interesting. Heibron quotes one contemporary observer as reckoning in 1762 that there were twenty mathematicians for every physicist, and quotes Lambert as saying in 1770 that 'For many years young people have emerged from universities knowing scarcely anything more than pure mathematics.' With such a training one could take up rational mechanics far more easily than experimental physics. As we have already seen, rational mechanics engaged the attention of every important mathematician of the century, including Clairaut, d'Alembert, and above all, Euler.

Heilbron, *Elements of Early Modern Physics*, pp. 9–10.

It would seem, then, that mathematics was at the centre of eighteenth-century science, or at least of the dominant part of that science, rational mechanics. Why was this so? The clearest statement of the view that this was a consequence of the natures of mathematics and science themselves was given by d'Alembert. He outlined his ideas in his *Traité de dynamique* of 1743, and again more thoroughly in his *Preliminary Discourse to the Encyclopedia* in 1751. Please *read an extract from the latter now* (**SB** 14.D3) and attempt the following.

Question 13

(i)   How, according to d'Alembert, are algebra, geometry, and mechanics related?

(ii) Why, then, did d'Alembert think that mathematics plays a central role in science?

Comment ————————————————————————————

(i)   D'Alembert argued that there is a chain of ideas. From the senses we learn of extension or space, filled by bodies (impenetrable shapes). Drop the idea of impenetrability and one has the idea of pure extended magnitudes, to which geometry is appropriate. By abstracting the physical and looking only at the rules for manipulating numbers (which might be the measurements of physical things) we arrive at algebra, the science of magnitudes in general. Or we can move in the other direction and pass from geometry to physics by enriching the conceptual mix.

(ii) Certainty, d'Alembert argued, is obtained by reasoning based on true and self-evident principles. Specifically, algebra is the most certain. Since geometry is a special case of algebra, algebraic certainty infuses into geometry up to the point where the idea of extension is opaque. Similarly, mechanics is a special case of geometry, made a little more obscure by the obscurities inherent in the concept of impenetrability. So mathematics, ideally algebra, is central because the other activities are special (and more obscure) cases of it, and the role of mathematics is to bring certainty to scientific thought. ∎

There is a markedly Cartesian flavor to this—d'Alembert is not much troubled by the interplay of measurement and analysis which occupied Clairaut. But d'Alembert's Cartesianism had an eighteenth-century sophistication. In contrast to earlier Cartesians like Malebranche, who readily fell back on *ad hoc* devices designed to keep alive some of Descartes' specific ideas (such as vortices), d'Alembert was a Cartesian only in clarity of spirit. He rejected mechanical artifices explicitly in favour of rigorous quantification and geometrical reasoning.

But d'Alembert's philosophy only fitted one kind of rational mechanics. D'Alembert always refused to ascribe physical reality to the force of gravity. For him gravity was an effect, the cause of which was not only unknown but also one not necessarily for the *scientist* to seek. When he wrote that 'the nature of movement is an enigma for the philosophers' he was quite prepared for it to remain their problem. In this he differed from many of his contemporaries: Pierre de Maupertuis, for example. Maupertuis moved more and more towards the position that gravity was a force which caused things to move. He organised the young generations of French Newtonians, apparently entertaining such as Voltaire, Algarotti, and du Châtelet to dinner on the days when the Académie des Sciences met, to which they would go full of 'good spirits, presumption, and strong arguments.' For them too, the force of attraction was a cause.

Quoted in Heilbron, *Elements of Early Modern Physics*, p. 50. Francesco Algarotti (1712–1764) was the author of the Italian book *Newtonianism for the Ladies*, 1737, a very successful popularisation of Newton's work and the inspiration for Voltaire's (see Figure 10).

Figure 16 Pierre de Maupertuis (1698–1759)

Figure 17 Alexis-Claude Clairaut (1713–1765)

The most original exponent of Newtonianism was Clairaut. When philosophical rigour demanded it he would, like Newton himself, claim that the concept of gravity did not explain anything because it was not understood. But writing to Euler he could be less guarded. The evidence of **SB** 14.B2–B3 suggests that he regarded attraction as real, questioning only the precise law by which it was to be described. This view came to be more and more widely held, and as it spread a certain philosophical laziness crept in: people began to feel that the effectiveness of Newtonian physics legitimised the theoretical constructs it contained. This differs from Newton's own mature views in that what Newton had treated mathematically but regarded as unexplained was now regarded as explained by the mathematics. Thus the simplest way to summarise the situation during the century would be to say that mathematics was generally taken to be the way nature was to be analysed and made to yield up its secrets. On this sophisticated and essentially Newtonian view, all scholars were agreed.

11.5 THE ANALYSIS OF NATURE

In the last section we looked at the position mathematics came to occupy in the study of rational mechanics. In this section we consider what effect this had on mathematics itself. We shall be interested in two particular consequences: the development of new mathematical techniques to solve problems arising in the study of nature; and the changing conception of mathematical analysis. We examine these consequences by considering three especially revealing examples.

Motion against resistance

Motion in a resisting medium may not sound a very inspiring topic, but it was critically important in its day, for it arose from the debates surrounding Cartesian physics. According to Descartes' vortex theory, the planets are carried round in an invisible medium, without which they would travel in straight lines. So part of Newton's purpose in discussing the topic was to analyse Descartes' idea, and as we saw (in Section 11.2) he was in fact able to refute it. There was another reason for studying the problem: the motion of a projectile travelling through the air is an important case of motion in a resisting medium, which the new mathematical techniques could deal with. Thus mathematicians do what anybody does with a new tool—they see what can be done with it. It was to be just so with the differential calculus.

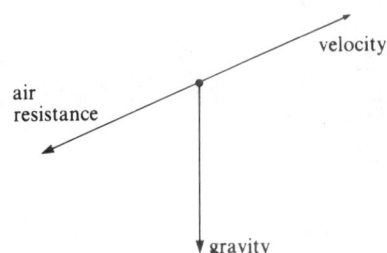

Figure 18   Forces on an object moving through the air

One of Leibniz' first observations on seeing a review of Newton's *Principia* in 1688 was that the (Leibnizian) differential calculus would be invaluable in building a theory of motion in a resisting medium. He at once published a paper along those lines, which made public some results although he suppressed his methods. The problem was a difficult one, which continued to interest him; in particular, he studied the case in which the medium exerts a resisting force proportional to the square of the velocity. Why, though, did he take up and persevere with this case of the problem (which was also the case Newton had considered in the *Principia*)? Like Newton, Leibniz thought the problem was an interesting one on physical grounds. If you imagine a solid body moving through a swarm of particles (perhaps a planetary vortex), then the resistance it will experience is proportional to the relative velocity of body and particle on impact, and also to the number of collisions in each unit of time. Since this number is proportional to the distance travelled, the total resistance is proportional to the square of the velocity.

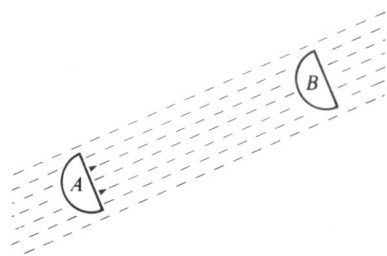

Figure 19   In time t the body moves with speed v from A to B, a distance vt. The number of particles it encounters is proportional to v, as is their force of impact

Leibniz discussed the problem with Huygens, before he died in 1695. Soon after Huygens' death another of Leibniz' correspondents took up the theme. This was Pierre Varignon (1654–1722). Varignon was an engaging man, who strove energetically to bring peace to the controversy over the invention of the calculus. He stood close enough to Newton to be able to commission Sir Godfrey Kneller to paint Newton's portrait for him in 1720. But he was also close enough to Leibniz and Johann I Bernoulli to be able to correspond steadily with them. Leibniz and he had been drawn together when, in 1700, the Académie des Sciences in Paris felt called upon to defend the calculus against the attacks of one Michel Rolle, who like Varignon was a salaried mathematician at the Académie itself. Rolle was an algebraist, well-trained in Cartesian techniques, who believed that the new Leibnizian infinitesimal calculus was not properly grounded and could lead to the wrong results. This was embarrassing for the Académie because Leibniz had become a foreign member earlier that year. As de l'Hôpital, the leading Parisian supporter of the calculus, was away, the Académie looked to Varignon; when he published his defence of the Leibnizian calculus Leibniz wrote to say how satisfied he was with it. Indeed Rolle some years later conceded the argument.

By this stage, Varignon had become quite expert at applying the differential calculus to problems of motion, and had shown how several of Newton's theorems about motion under a centripetal force could be derived in this way. He could show, for example, how to determine the law of force given the orbit, at least in simple cases. Leibniz wrote to encourage him in his endeavours, and to suggest that he try this more interesting problem (the converse one): given several bodies attracting one another, find their orbits. This problem proved intractable because determining the

Figure 20 This now badly deteriorated portrait of Sir Isaac Newton at the age of seventy-seven, was painted by Kneller for Pierre Varignon in 1720

Figure 21 Pierre Varignon (1654–1722)

orbit given the law of force requires at least a grasp of the integral calculus—a subject not then known to Varignon. Moreover, his request to Bernoulli for instruction was turned down as a consequence of the contract Bernoulli had signed with de l'Hôpital. Nor could he have learned to integrate from Leibniz' paper on motion in a resisting medium, for Leibniz had not revealed his methods there. However, over the next few years Varignon patiently caught up, and did much to show how the calculus could solve many problems to do with motion. Specifically, he showed how the formalism of differential equations could be used to advantage in the study of dynamics, presenting a general technique in place of Newton's ingenious but *ad hoc* geometrical arguments.

As we saw in *Unit 10*, Section 3.

The culmination of this line of approach was Euler's early book on mechanics (*Mechanica*, 1736), where he discussed many cases of point-masses under various laws of force and subject to various forms of resistance. The result was a book which showed how to solve many differential equations and how to obtain the curves that they describe. Still, one might have reasonably wondered what this had to do with the subject of mechanics! Newton had shown that it was permissible to treat planets as point-masses so far as their gravitational interaction was concerned, and one might imagine other problems where it was a reasonable approximation to consider a finite body as a point. But there would be many situations and problems where this assumption would be unsatisfactory. Indeed, in the book Euler outlined a six-part programme for mechanics. He suggested that the following topics be taken up:

(i) infinitesimally small bodies;

(ii) finite rigid bodies;

(iii) flexible bodies;

(iv) expansion;

(v) several disjoint bodies;

(vi) fluids.

Not even Euler could work through this programme entirely, but he and his contemporaries did work on most of it over the next forty years.

Euler's initial steps down this road, then, resulted in a book on mechanics that looks more like a book on the geometry of curves. It is worth emphasising this and considering why it had this appearance. The most likely reason is the naive-sounding one that this was all that Euler, or anyone else for that matter, could do at the time. At any stage the shape of a subject is constrained by the techniques available to study it. In this case it would seem that the ability to formulate and solve mechanical problems in the language of the calculus was restricted to problems of this geometrical kind. That said, Euler's work did establish how much the calculus could already do.

Orbits and laws of motion

The calculus had also gained a significant success when Johann I Bernoulli and Jakob Hermann independently tackled a problem that Newton had rather skated over in his *Principia*. This was the determination of the orbit of a planet under the assumption that it is subject to an attractive force obeying an inverse square law. Newton had showed how to find the orbit, so when Bernoulli and Hermann took up the problem it was not, strictly speaking, unsolved, but Newton's solution was written in the geometrical style of the *Principia*, and Bernoulli seems to have felt that it lacked something in generality and directness.

The story is an entertaining one. At this time—the years around 1710—Newton was engaged with Roger Cotes in producing the second edition of the *Principia*. Nikolaus Bernoulli, Johann's nephew, came to London in October 1712, and told Newton that there was a mistake in one of his theorems concerning motion through a medium that resists in proportion to the square of the velocity. It seems that this was a discovery of Johann's in 1709, and being unable to see where Newton had gone wrong he had produced a better proof of his own. But he also convinced himself (wrongly) that the mistake was evidence that Newton did not understand the subtleties of the calculus—with, if true, an obvious bearing on the priority dispute. At the same time he decided to put forward his alternative to Newton's treatment of the inverse square law. If you look at his discussion of this point in **SB** 13.B3, you will see just how firmly Bernoulli then based his own account of this problem on the calculus, and also how pleased he was to surpass Newton, as he saw it. It seems that he then tried to arrange for his two papers to come out just after the second edition of the *Principia* was to appear, thereby correcting these mistakes publicly. However, the plan went wrong. The edition was delayed, and his nephew Nikolaus told Newton of the error. Newton, believing Johann Bernoulli to have acted out of generosity, proposed him as a foreign member of the Royal Society. When he later saw the trap that had been prepared for him, he was less pleased.

Nonetheless, Bernoulli's achievement was a real one, for he showed that calculus did indeed simplify arguments that could be carried out in other ways only with difficulty, and his suspicion that the calculus would be essential to making progress was to be amply confirmed. This is a good example of how throughout the first half of the eighteenth century many mathematicians worked hard to understand the *Principia*, and by their attempts to clarify and understand it were led to increasing use of the calculus. The consummation of this process is perhaps best captured by a remarkable paper of Euler's, written in 1750.

Question 14   Please look now at **SB** 14.C1(a), and the commentary on it by Truesdell (**SB** 14.C1(b)). What, in general terms, is Euler presenting here?

Comment

Euler presents three formulae which describe the motion of a point-mass completely; he speaks of the formulae as containing 'all the principles of mechanics'. More than that, they are derived in a simple, uniform way by means of the calculus. Henceforth, it was possible to write them down at the start of a problem, rather than work out from scratch in each case just what Newton's laws of motion were trying to say.   ■

If you have studied mechanics, you may be surprised to see that it was Euler and not Newton who first wrote down the laws of motion in this form. You may also have found Truesdell's account rather condensed, plucked as it was from a very rich analysis, so let us discuss it further here. Truesdell argues that one of the crucial things Euler did was to introduce rectangular Cartesian axes as standard. This is just what he did when analysing curves in his *Introductio in analysin infinitorum*.

See **SB** 14.A4 and *Unit 12*, Section 2.

In the present context, Euler explained in the third paragraph of **SB** 14.C1(a) how to use fixed perpendicular axes. It was a bold move; the tradition going back at least as far as Descartes had been to use coordinate axes which arose out of the special

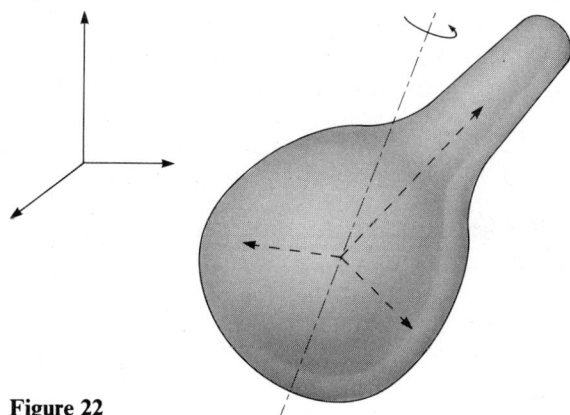

Figure 22

needs of the problem at hand, and were not necessarily right-angled or laid down in advance. Such coordinate systems are what Truesdell calls 'intrinsic', and their advantage is that they can cut down significantly on the amount of calculation involved in working out a particular problem. Euler's insight was that the cost in increased complexity from introducing rectangular axes would be more than outweighed by the gain in conceptual simplicity. The gain shows up in two ways. One can develop a theory which can be used easily in a range of problems, precisely because the coordinate system has been standardised. And the notation itself makes it obvious that one can add quantities which have both magnitude and direction (*vectors*), thus bringing into the domain of mathematics quantities which had hitherto been difficult to handle.

Newton's conceptions of velocity, acceleration and force (Box 1) were vectorial in this sense.

This step forward, although simple, was important. For example, one reason why Euler's analysis of the motion of the moon finally surpassed Clairaut's was that it was couched throughout in this system of fixed rectangular coordinates. Euler achieved similar success with his analysis of the rotation of a solid body, which he carried out in 1758. Here he showed,

(1) that each body could be thought of as carrying a set of perpendicular reference axes, and

(2) how these axes were moving at any instant with respect to a set of axes fixed in space.

This was a considerable enlargement in the scope of mechanics, amounting to the second stage of Euler's six-part programme of 1736. Since the third stage, flexible bodies, was well underway with the investigation of the vibrating string, and the sixth stage, fluid dynamics, was also making good progress (although we have not been able to tell that story) it is clear that a considerable broadening of the subject was taking place.

Variational principles

For our last illustration of how new mathematical techniques and concepts solved problems in the eighteenth-century study of nature, we turn to the growing use of *variational principles*. In the middle of the previous century, Pierre Fermat had put forward the archetypal example of such a principle when he argued that it was a property of light that it always took the least possible time to traverse a journey. In this way the study of optics was to be advanced by a mathematical analysis of all possible routes light *could* take; the route it *would* take was that traversed in shortest time. In due course the type of mathematical analysis needed to find the minimum was to be decisively advanced by new techniques in the differential calculus, but let us first consider some examples of Fermat's *principle of least time*, to become clear about the idea, before looking at those later developments.

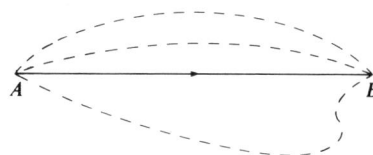

Figure 23

As the simplest example, what determines the path taken by light in going from one point, A, to another, B, in a uniform medium (see Figure 23)? Any route other than the straight line AB would take longer to traverse. So AB is the path it takes.

Again, why does light reflect off mirrors in the way it does (making equal angles with the mirror surface)? Because if it reflected off any other point the total path would be longer, and so the journey time would be greater.

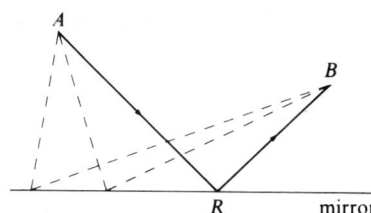

Figure 24 The route *ARB* is shorter than any alternative reflection, as was known to Heron (see **SB** 5.A3(c))

The case Fermat considered in detail was that of light travelling through two different mediums, say air and water, where the speed through each will be different. Under these circumstances it will not travel along the straight line joining A and B but along a path which takes less time. This explains the phenomenon of *refraction*.

As it happens, it was Fermat's view that light should travel through a dense medium more slowly than through a rare one, whereas Descartes held the contradictory (and in some views, self-contradictory) position, so Fermat was surprised to discover that his analysis confirmed Descartes' law of refraction. In a move that foreshadowed how things were to go, Fermat's analysis of the problem of refraction used a method of maxima and minima to find the actual path of the light.

In each of these cases, if one asks for the path of the light given such-and-such information about the medium it is to pass through, the answer will be a (straight or curved) line. There are other problems of this kind which people began to ask. A famous one concerns the *brachistochrone*. This is defined to be the curve of quickest descent joining any two of its points. So the problem is to discover what *is* this curve—is it a straight line, or some other known curve, or a hitherto unknown one? To find the brachistochrone joining A and B one must therefore consider *all* the curves joining A to B. Imagine rolling a ball down each one and timing its descent: the brachistochrone is the curve for which the journey-time is least. It comes as something of a shock to realise that the winning curve is not a straight line, and in the 1690s Johann Bernoulli organised a competition to find out what it was. The answer turns out to be the *cycloid* through A and B which has a vertical tangent at A. (So in this race it turns out to be vital to get off to a quick start!)

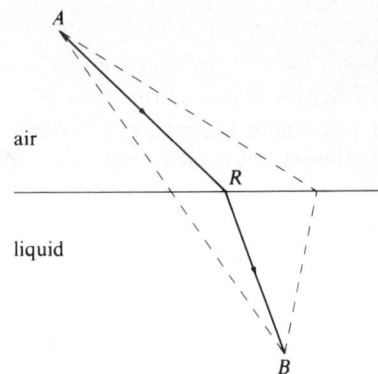

Figure 25 The route *ARB* takes a shorter time than the straight route *AB*

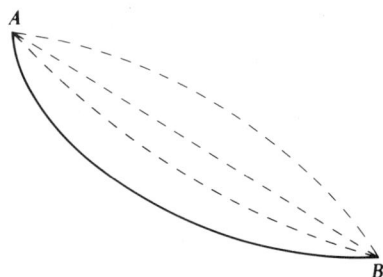

Figure 26(a) Which is the quickest path of descent from *A* to *B*?

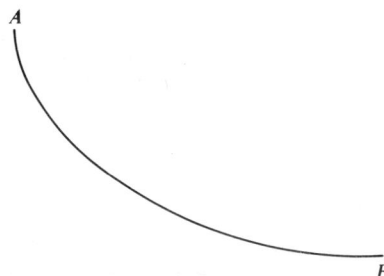

Figure 26(b) The brachistochrone (a cycloid)

Interest in problems where the solution is a curve which minimises something was quickened by two theoretical developments in the middle of the eighteenth century.

First, various mathematicians were able to show, drawing on ideas of Leibniz, that in a whole range of physical problems a particular quantity (much later to be called the *energy* of the system) was conserved. For example, when a body falls it gains in speed as it loses height. When this is put into theoretical terms, the gain in speed is a gain in *kinetic energy*, the loss of height a loss of *potential energy*, and the sum *kinetic energy + potential energy* turns out to be constant. The total energy involved when two bodies collide is likewise a constant. This principle of conservation turned out to be a very elegant way to formulate a mathematical theory of mechanics. It is not quite the same as Fermat's principle, which required something to be minimised, but it shares with that principle the idea that, to understand how a physical system changes in time, one looks at all the mathematically possible ways and selects the one satisfying some simple requirement which can be stated in advance (something is minimised, something is conserved).

The second development is actually a synthesis of Fermat's and Leibniz' ideas. Maupertuis, who (by 1745) was in charge of the Berlin Academy of Sciences, proposed that all the laws of motion and mechanics could be deduced from a single principle—which he called the *principle of least action*. We shall not attempt to define what he called the 'action' of a physical system, not least because his own attempts to define it were all rather obscure and even inconsistent. It is sufficient to notice his claim, that to understand how a system behaved it would be enough to analyse it on the assumption that a certain quantity was kept to a minimum.

Unluckily for Maupertuis, his principle was badly received. He had been joined in Berlin by Voltaire and a Leibnizian he had known for many years, Samuel König. But König attacked the principle itself, claiming that in any case what was good in it was due to Leibniz, and Voltaire attacked Maupertuis' deduction from it—that God must necessarily exist—in a particularly vicious satire. Frederick the Great, who took a close interest in his Academy, sided with Maupertuis, and passions became so heated that Voltaire had to leave for his safety and Maupertuis for his health. Once matters had calmed down Euler was able to salvage some of Maupertuis' principle; he could define precisely the action of a system in a certain range of cases, something Maupertuis had never done. Then in 1761, Lagrange was able to propose a quite general principle of least action for systems of point-masses. For the first time a fairly large set of mechanical concepts was put into order without invoking the concept of force—Leibniz and Huygens would have been pleased, one suspects, at this surprising vindication of their belief that a force-free mathematical physics should be possible.

Diatribe du Docteur Akakia (1752).

Finally we reach Lagrange's *Méchanique analitique* (*Analytical mechanics*) of 1788. This book is remarkable in many ways. It bases the whole study of mechanics on the idea of variational principles; and it is thoroughly algebraic. As Lagrange said proudly in its introduction,

> There are no figures in this book. The methods that I demonstrate here require neither constructions, nor geometrical or mechanical reasoning, but only algebraic operations, subject to a regular and uniform development.

Quoted in T. Hankins, *Science and the Enlightenment* (Cambridge University Press, 1985) p. 29.

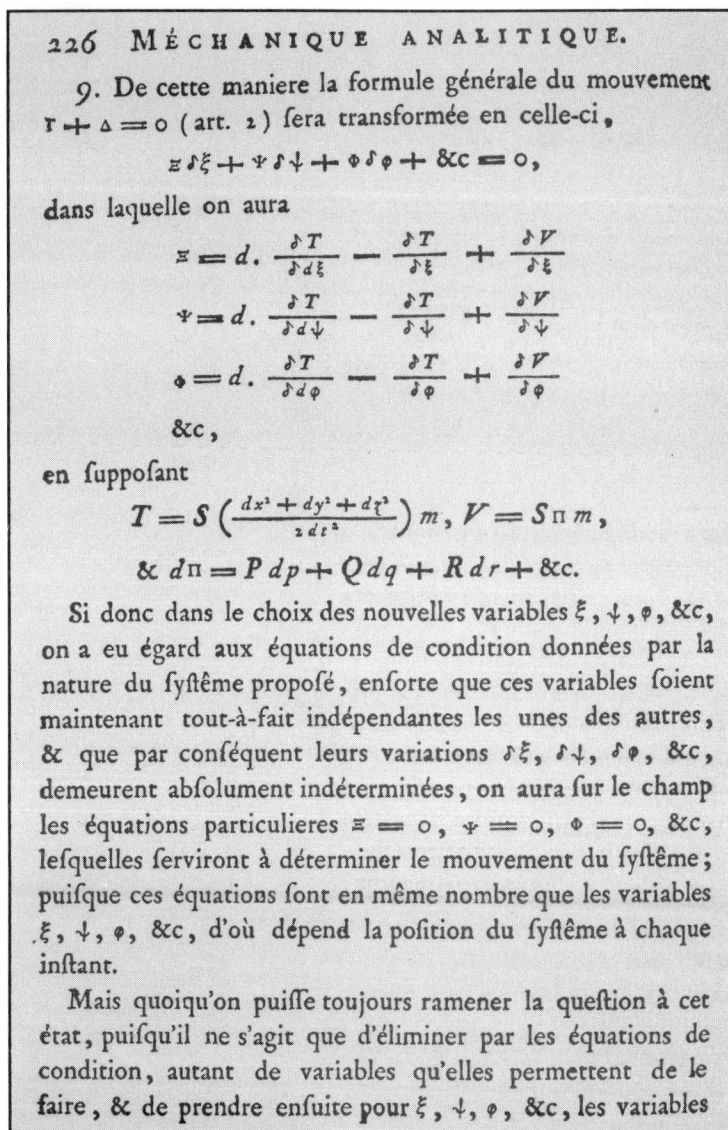

Figure 27 A page from Joseph-Louis Lagrange's *Analytical mechanics* (1788), showing 'no figures ... but only algebraic operations'

Lagrange was indeed the master of variational problems, which he elevated into a new field of mathematics, the *variational calculus* (as Euler christened it). In *Méchanique analitique* the theory of mechanics was rewritten in the language of the calculus. Views are divided about the value of Lagrange's achievement here. Despite its undeniable mathematical elegance, some have argued that it was much better at making sense of what was already known than helping to discover new properties of nature. Truesdell has observed that unlike the work of Euler, or even Newton's *Principia*, in Lagrange's *Méchanique analitique*,

> nothing [looks] forward rather than backward, and [there is] no comparison of theoretical and observed values. Mechanics appears as a physically closed subject, a branch of the theory of differential equations.

Truesdell, 'A program toward rediscovering the rational mechanics of the Age of Reason', p. 34.

We conclude by considering what has been meant by the phrase 'the *analysis* of nature'. Analysis originally—as it was defined by Pappus, for example—meant the taking apart of something, much as chemists use the term today. In the hands of Viète and Descartes it became inextricably algebraic, a process that was accelerated with the introduction of the calculus, for it was the formal side of the calculus that proved to be so useful in the study of problems of motion. To analyse a problem in mechanics became to cast it in the language of the calculus, and ideally to solve it there. The more acceptable it became to leave the answer in the form of an expression or a function, the more the process of solving a problem stopped when the analysis of it stopped. Since the calculus had its own logic and consistency, it was appropriate to base one's study of mechanics on it, and as we have seen, that is what the mathematicians of the eighteenth century did. First Newton's work was re-done in this way; we saw how Johann Bernoulli, Varignon, and Euler brought in more and more of the machinery of differential equations to tackle problems of motion. Then, with the use of the calculus of variations, the subject was re-formulated in a way that could elucidate its most basic theoretical insights only by applying the most advanced techniques of the calculus. Rational mechanics and mathematical analysis became completely fused in this way. So much so, indeed, that the nineteenth century was to find that further progress could only be made by breaking this union apart. Rational mechanics was swamped by the successful quantification of the domains of heat, electricity and magnetism, which were found to be markedly non-Newtonian in their theoretical presuppositions, and by the move towards a more experimentally-led investigation of nature. And finally the calculus was to be taken out of the grip of formal algebra and based on more secure foundations, as you will see in *Unit 16*. All that then remained of its origins as an aid to thought was its name: to distinguish it from its earlier forms the rigorous calculus has become known as mathematical analysis.

FURTHER READING

Cohen, I. B., *The Newtonian Revolution* (Cambridge University Press, 1980).
This book combines a detailed look at *Principia* with a thesis about revolutions in scientific thought. It is not always an easy read, but is unfailingly thought-provoking, and a good companion if you ever want to sit down with the *Principia* itself.

Hankins, Thomas L., *Science and the Enlightenment* (Cambridge University Press, 1985).
A good brief introduction to science in the eighteenth century, especially attentive to the French viewpoint. It covers not only mathematics but the experimental sciences and natural history as well as the moral sciences. Not unduly technical, and rich in the interconnections it makes, this study also contains an extensive and useful bibliographic essay.

Heilbron, J. L., *Elements of Early Modern Physics* (University of California Press, 1982).
The best recent study of physics (as opposed to rational mechanics) in the seventeenth and eighteenth centuries. It is rich in depth, although narrower and sometimes more technical than Hankins' book.